Connection on the Ice

In the series *Environmental Ethics, Values, and Policy,* edited by Holmes Rolston, III

Connection on the Ice

Environmental Ethics in Theory and Practice

Patti H. Clayton

With photographs by Charles Mason

Temple University Press

Philadelphia

Temple University Press, Philadelphia 19122
Copyright © 1998 by Temple University
All rights reserved
Published 1998

Printed in the United States of America

Text design by Erin Kirk New

♾ The paper used in this publication meets the requirements of American National Standard for Information Sciences—Permanence of Paper for Printed Library Materials, ANSI Z39.48-1984

Library of Congress Cataloging-in-Publication Data
Clayton, Patti H.
  Connection on the ice : environmental ethics in theory and practice / Patti H. Clayton ; with photographs by Charles Mason.
    p.   cm.—(Environmental ethics, values, and policy)
    Includes index.
    ISBN 1-56639-615-8 (alk. paper).
    ISBN 1-56639-616-6 (pbk. : alk. paper)
    1. Environmental ethics.   2. Environmental sciences—Philosophy.
  3. Wildlife rescue.   4. Whales.   I. Title.   II. Series.
GE42.C55   1998   94-41265

Epigraph by Christopher Stone from "Should Trees Have Standing?—Toward Legal Rights for Natural Objects," in *People, Penguins, and Plastic Trees: Basic Issues in Environmental Ethics,* ed. Donald VanDeVeer and Christine Pierce (Belmont, CA: Wadsworth Publishing Co., 1986), 92.

Frontispiece: The experience of human–nonhuman connection. In a defining moment of what was perhaps the "most extraordinary animal rescue effort ever undertaken," Barrow, Alaska, residents reach to touch one of three ice-entrapped gray whales. © 1988 Charles Mason.

To my family,

    Princequillo,

    Chrissie,

    and Kevin,

for the gift of "living in a moment"

What is it within us that gives us this need not just to satisfy basic biological wants, but to extend our wills over things, to object-ify them, to make them ours, to manipulate them, to keep them at a psychic distance? Can it all be explained on "rational" bases? . . .

To be able to get away from the view that Nature is a collection of useful senseless objects is . . . deeply involved in the development of our abilities to love— or, if that is putting it too strongly, to be able to reach a heightened awareness of our own, and others' capacities in their mutual interplay. To do so we have to give up some psychic investment in our sense of separateness and specialness in the universe. . . . Yet, in doing so, we—as persons— gradually free ourselves of needs for supportive illusions.

—Christopher Stone

Contents

Acknowledgments

An author's first book, this one anyway, is the culmination of a lifetime of study and experiences and relationships, and so the opportunity to express appreciation to all of those who have influenced and supported its development is a welcome, albeit challenging, one. I trust that all of those family members and friends, students and teachers, mentors and colleagues who are not mentioned by name will acknowledge, at least in their own hearts, the gifts they have given and the gratitude I feel.

It goes without saying that this study could not have been successfully completed without the participation of interviewees and others who contributed to the process of collecting data on the Barrow whale rescue. Many people in Alaska and elsewhere gave generously of their time, pointed me in important directions, allowed me access to their personal files, and openly shared their memories of this remarkable event and their insights into the "hows" and "whys" of its unfolding; the richness and detail of this study are the direct result of their thoughtful and candid cooperation. The primary participants in the study are listed in the Appendix; while this is undoubtedly not the final product any of them envisioned, it owes its very existence to them and I cannot thank them enough for their interest in the project and for their willingness to return to this story once again. Charles Mason is owed my appreciation several times over for the photographs that so beautifully grace this text and so powerfully transport its readers to the very side of the whales. I struggled diligently to get all the details straight and to keep the various versions of the story recounted herein true to their commentary, and any factual errors, misunderstandings, or misinterpretations are my own.

I have been fortunate to have an extraordinarily supportive group of colleagues at the University of North Carolina–Chapel Hill (UNC) and at North Carolina State University (NCSU). Individually and collectively, Pete Andrews, Seth Reice, Bruce Winterhalder, Catherine Marshall, and Anthony Weston encouraged me to stretch the already malleable boundaries of the UNC Ecology Curriculum and to engage in nontraditional and creative research at the intersection of natural science, social science, and philosophy. I am most grateful for this flexibility, for the host of unique teaching and learning opportunities they provided, and for their often excessive confidence in my abilities. And my colleagues at NCSU, especially Erin Malloy-Hanley and Patrick Hamlett, provided invaluable guidance through my early years as an instructor, through some of the relevant philosophical literature, and through the publication process itself. My teachers, my fellow faculty members, and my students at both universities are the source of years of inspiration, course correction, and encouragement, and, again, any shortcomings of this study are mine not theirs.

As my first foray into the realm of publishing, this text benefited significantly from the efforts of everyone at Temple University Press, from the comments of the anonymous reviewers affiliated with both Temple and the MIT Press, and from the feedback of students at UNC–Chapel Hill and Elon College (who admirably met the unenviable task of making their way through its first, heftier incarnation). A special thank you to Doris Braendel for her patient guidance and editorial expertise and to Holmes Rolston for his support and his role as liaison.

And finally, my deepest appreciation is due to my family, who provided both direct assistance during the course of this study and a lifetime of encouragement and support: my parents Hassell Hilliard Sr. and Pat Hilliard (who eagerly and quickly transcribed more than three hundred pages of interviews) and my brother Hassell Hilliard Jr. (who housesat while my husband and I were in Alaska conducting the research). Ruth Clayton also helped to take care of our feline family, and Ed Clayton provided computer support; they, and the rest of my "new" family, have also been a constant source of support.

Closest to home, and to my heart, Kevin, Princequillo, and Chrissie—as well as "our" birds, squirrels, and "bear-faces"—provided both the inspiration and the motivation for my work. In many ways, the final product is as much Kevin's as my own, thanks to the shared adventure in Alaska, the endless hours of formatting and reformatting, the gift of time, and the esoteric and inevitably late-night discussions of philosophy. This text began and was nurtured during the "best of times" in our life together, and if it reached completion during the "worst of times" it will nevertheless stand as a reminder of the deep joy that comes with deep connection, truly known and cherished.

This dramatic rescue effort has captured the world's attention . . . and perhaps reminded all of us of something essential about ourselves and our human nature.

Introduction

In *The Call of Stories: Teaching and the Moral Imagination* Robert Coles writes of the "immediacy that a story can possess, as it connects so persuasively with human experience," and he encourages "respect for narrative as everyone's rock-bottom capacity, [and] also as the universal gift to be shared with others."[1] Long admired as an artform, storytelling and the study of personal and communal narratives are becoming something of a science, as Coles's psychiatric work, among a host of examples in a variety of disciplines, attests. Seriously attending to stories as meaningful "windows" on reality as it is perceived, experienced, and constructed is one of the defining methods of qualitative research and one of the central components of the emerging "postmodern" western worldview. This study itself is an outgrowth of, and hopefully a contribution to, the emerging science of storytelling.

Here we explore in depth one particular story of human–nonhuman encounter: the 1988 rescue of ice-entrapped gray whales near Barrow, Alaska, an event which has become part of our national story, part of the cultural story of Alaska, part of the life stories of the rescuers, and, perhaps, part of the family story of certain gray whales. This study attends to the varied stories recounted by participants in the rescue as they attempt to make sense of and learn from the encounter; and in turn it presents new stories that serve as vehicles for the study of other such episodes of human–nonhuman interaction. It seeks to help establish a space within environmental philosophy for such storytelling. And, of course, the study has its own story, which we turn to first in the hope of shedding light on its evolution, objectives, structure, and method.

In many ways, the story of *Connection on the Ice* began in the mid-1980s. By the time the story of the whales was capturing the attention of the world, I had already been immersed for several years in interdisciplinary environmental studies. My interests lie in what is sometimes called the "human–environment problematique": that area of study and practice in which we struggle to understand and sometimes alter the myriad interactions between human beings and the natural world and its nonhuman inhabitants—interactions that range from individual relationships with nonhuman animals to interhuman conflicts over the form and extent of resource use, and from international policy debates on sustainable development to individual consumer choices in the marketplace. A focus on such issues has served as my guiding theme through an exciting foray into a wide range of disciplines, including, among others, literature, biology, policy, ecology, history, anthropology, and, finally, philosophy.

The turn to philosophy occurred as a result of frustration with the conceptualizations of human–environment issues offered in most of these other disciplines: in none of them did I find an expression of my own growing sense that all of these issues were essentially products of one core (but in my own mind unnamed) dynamic. Before looking to philosophy for further clarification, the closest I was able to come to pinning down this elusive inadequacy was that none of the other disciplines truly addressed the meaning or significance of the fundamental relationship between ourselves as human beings and the other members of the natural world or what it might mean to live a fulfilled human existence in that context. Rather, it seemed that each in its own way rendered the nonhuman aspect of this problematique (and all too often the human aspect as well) as a mere "thing" or "object" that had little more than instrumental value. Surely moral philosophy, particularly in the form of environmental ethics, held the key to a different sort of conceptualization of these issues: one that acknowledged the fundamental role of human–human and human–nonhuman relationships, both conceptually and practically, in these issues.

A grant from The Pew Charitable Trusts' Initiative for Integrated Approaches to Conservation and Development allowed me to spend a year of intensive study of philosophy and environmental ethics, exploring how we can conceptualize issues of conservation and of development, not as separate entities, but as one problematique bound together by their mutual grounding in the ethical relationship between self and "other," be the "other" human or nonhuman, system or individual, present or future. But the same sense of incompleteness soon set in once again. Although it could indeed recast conservation and development issues in terms of their common ethical underpinnings (basing obligation, for example, on respect for inherent value, whether of starving children, of endangered African elephants, or of future generations), the philosophical canon—grounded in the same tradition of rationalism as most of the other disciplines—appeared to embody a similar sense of objectification, alienation, or distancing of self and other.

And thus I was finally able to pin down that elusive underlying dynamic I had searched for as the common thread uniting the myriad issues of the human–environment problematique: our failure to understand ourselves as fundamentally in relationship with, or connected to, the "objects" of our concern (again, be they human or nonhuman) in many ways underlies both our environmental and social failures and the pervasive experience of our lives as unsatisfyingly alienated and unfulfilled. Vice President (then Senator) Al Gore, author of *Earth in the Balance,* made the point much more simply and eloquently:

> Just as the false assumption that we are not connected to the Earth has led to the ecological crisis, so the equally false assumption that we are not connected to each other has led to our social crisis. Even worse, the evil and mistaken assumption that we have no connection to those generations preceding us or those who will follow us has led to the crisis of values we face today. Those are the connections that are missing . . . today; those are the bridges we must rebuild . . . and those are the values we must honor if we are to recapture faith in the future.[2]

After I had looked in vain for this conceptualization of the human–environment problematique in the philosophical literature, a colleague suggested I move beyond the mainstream philosophical traditions I had been exploring and pointed me toward feminist philosophy, and I recalled an earlier brief exposure to the writings of Martin Heidegger. Success at last! Within these two traditions I did indeed find a less-alienated conceptualization of the relationship between self and other. The "care" tradition within feminist philosophy and the phenomenological work of Heidegger both posit a self fundamentally "connected" with others, and the environmental philosophy "offspring" of these two frameworks, unlike that of the dominant philosophical tradition, adopt "connection" rather than detachment as the fundamental premise of human–nonhuman interaction.

At this point, the objectives of and structure for this study became clear. The mainstream tradition, the one that environmental ethicists quite reasonably have turned to first in the development of this new field, seems to be in need of supplementation from alternative perspectives, two of which are considered here. And the judgment that the dominant and alternative frameworks bring fundamentally different perspectives to environmental ethics questions suggests their function as "conceptual lenses": the way we perceive our relationship with nonhumans and the natural world at large, and thus the type of theory building we engage in, may well be affected by the "lenses" we adopt. Comparing and contrasting the influence of philosophical traditions as conceptual lenses is best facilitated by grounding the critical comparison perspective in analysis of a single event; the more intriguing and rich the event, the better, and few well-known environmental ethics events of recent years hold the appeal of the gray whale rescue of 1988. Thus, the present study has three central objectives.

(1) To explore the influence of philosophical traditions as conceptual lenses that shape our thinking about environmental ethics issues

> Answers to questions like why the Soviet Union tried to sneak strategic offensive missiles into Cuba must be affected by basic assumptions we make. But what kinds of assumptions do we tend to make? How do these assumptions channel our thinking? What alternative perspectives are available? This study identifies the basic frame of reference used by most people when thinking about foreign affairs. Furthermore, it outlines two alternative frameworks. Each frame of reference is, in effect, a 'conceptual lens.' By comparing and contrasting the three frameworks, we see what each magnifies, highlights, and reveals as well as what each blurs or neglects. . . . By addressing central issues of the crisis first from one perspective, then from a second, and finally from a third, [this study] not only probe[s] more deeply into the event, uncovering additional insights; [it] also demonstrate[s] how alternative conceptual lenses lead one to see, emphasize, and worry about quite different aspects of events like the missile crisis.[3]

If imitation is indeed the highest form of flattery, then Graham Allison—from whose book, *Essence of Decision: Explaining the Cuban Missile Crisis,* this quote is taken—should be quite flattered by this study, for it is explicitly modeled on his own: substitute "whale rescue" for "missile crisis" and "environmental ethics" for "foreign affairs" and we have this work's first objective in a nutshell. Just as Allison's text explores the role of three different theoretical frameworks in political science as conceptual lenses that shape our understanding of events like the Cuban missile crisis, this study will explore the role of three different theoretical frameworks in moral philosophy as conceptual lenses that shape our thinking about environmental ethics events.

Identifying the dominant framework in moral philosophy and suggesting two alternative frameworks, we will explore the development of each body of theory in the realm of environmental ethics, and we will consider how the assumptions underlying each shape our understanding of the event in question: the whale rescue. As we will see, each framework, or lens—based, as Allison notes, on often radically different "assumptions, categories, and angles of vision"—offers a unique understanding of ethical deliberation and sheds new light on human–nonhuman relationships. We will thus compare and contrast the three frameworks with respect to the elements of the event they variously highlight and neglect, and we will explore, through three different recountings, how each yields a unique understanding of the whale rescue.

(2) *To assess the adequacy or completeness of the dominant theoretical framework in environmental ethics and to explore the potential of two alternative frameworks as supplements*

> Morality confronts us as an alien set of demands, distant and disconnected from our actual concerns . . . and brings with it alienation—from one's personal commitments, from one's feelings or sentiments, from other people, or even from morality itself.[4]

> Only [the] experience of connectedness will save the earth—and us with it. Any attempt, however grandiose and with however much commitment to its cause, will fall short if it does not have at its root this transformation of human experience in which human thinking knows connectedness as such and itself within that.[5]

This study is intended to be a contribution to the ongoing dialogue between dominant and alternative theoretical frameworks in the discipline of moral philosophy, particularly as it involves the emerging field of environmental philosophy. The subtitle, "Environmental Ethics in Theory and Practice," not only alludes to the combination of theoretical and empirical analysis of environmental ethics questions characterizing the study; it also suggests a discrepancy between the theory and the practice of environmental ethics, a discrepancy mirrored in the contrast between the dominant and alternative frameworks under investigation. The pairing of the concepts of "alienation" and "connection" in the preceding quotations brings this discrepancy into sharp relief: it is a central thesis of this study that mainstream environmental ethics has more to do with the former term (Railton's criticism of "morality" is directed toward the dominant tradition), whereas the actual experience of moral agency—of people making decisions about nonhumans and the natural world in general—may be more closely associated with the latter term (and thus more in line with either or both of the alternative conceptual lenses, which were selected as potential supplements because of their explicit assumption of connection between self and other). The two alternative frameworks under consideration here speak to the human–environment relationship and to the process of ethical deliberation in that context, but without the dominant tradition's limiting assumptions of detachment between self and other, and thus they may shed meaningful light on the actual practice of environmental ethics decision-making.

If it is indeed the case that our sense (or lack thereof) of relationship with or connection to nonhuman "others," rather than the inherently alienating adoption of the detached "moral point of view" posited in the dominant tradition, best captures the essence of moral agency (the process of decision-making, the way we perceive ourselves as agents, the way we understand and relate to the subjects of our concern, and the nature of the judgments we reach regarding our treatment of them), then our alternative frameworks do indeed serve as necessary supplements to the dominant framework. If this discrepancy between moral agency as it is conceptualized in the literature and as it is actually experienced in our everyday lives (or in extraordinary instances such as the whale rescue) does indeed exist, then it seems to follow that much mainstream environmental ethics theorizing is at best incomplete and at worst irrelevant . . . and in any case in need of supplementation from these other perspectives. The second objective of this study, then, is to assess the completeness of the dominant environmental ethics framework in this light; this will involve making implicit assumptions in each of the three traditions explicit, comparing and contrasting the relative strengths and weaknesses of the three lenses in making sense of the event, and otherwise evaluating the three frameworks in a critical comparison mode.

(3) To gain a deeper understanding of the whale rescue as an environmental ethics issue and as a microcosm of the human–environment interaction and to assess the value of real-world grounding for environmental ethics theory building

This dramatic rescue effort has captured the world's attention . . . and perhaps reminded all of us of something essential about ourselves and our human nature.

Does the profession of moral philosophy now display that degeneration . . . where there are beautiful souls doing their theoretical thing and averting their eyes from what is happening in the real world . . . ? . . . The retreat of the theorists from moral problems in the real world to the construction of private fantasy moral worlds . . . [has displaced the vision of] the relation between moral philosophy and the actual human practices in which appeals to moral judgements are made and in which morality makes a difference to what is done, thought, and felt.[6]

It is with this concern for the relevance of theory to actual experience that the rescue of gray whales enters the picture. Assuming that the stories people tell of their ethical decision-making and their values have at least as much relevance as do theory-driven models in the effort to better understand the experience of human moral agency, we will take the somewhat nontraditional (for philosophy, that is) approach of grounding the exploration of dominant and alternative frameworks in a real-world case study. A corollary aspect of the theoretical critique, then, involves the legitimacy of grounding the development of theory in the actual experience of moral agents; if the discipline is to have any claim to relevance and meaning in our lives, philosophers must begin talking to real people making real choices, and our third objective is pursued in this spirit.

The case study of interest here is the October 1988 rescue of gray whales trapped in ice off Barrow, Alaska. Described in detail in Chapter 1, this event embodies many of the controversies, complexities, and values that characterize most environmental ethics issues, and it highlights the decision-making processes and experiences of a wide range of individual participants and observers. A three-week rescue effort that cost over $1 million, it has been hailed as one of the most remarkable animal rescues in history. It ultimately involved the White House, the (then) Soviet Union, the environmental community, Eskimo whalers, the National Oceanic and Atmospheric Administration (NOAA), the Alaska National Guard, Alaskan oil corporations, and companies, media representatives, school groups, and families throughout the United States and around the world. It has been criticized for its expenditure of resources on animals in a world wracked by the illiteracy and malnutrition of countless members of our own species. It has been dismissed as a media- and public-relations-driven farce. And it has been lauded as one of our finest hours, marked by cooperation among ideological enemies, by self-sacrifice, and by rarely displayed empathy with the suffering of other beings. What makes it most interesting is that it is all of these things and more.

The study's third objective, then, is to explore this intriguing story—the story of three gray whales, of the people who spent three amazing weeks at their side, and of the attention the entire world focused on this unprecedented rescue effort—from the perspective of the participants in and observers of the event as well as from that of each of our three theoretical frameworks. In the process, we will consider the lessons this particular environmental ethics issue may hold for our understanding of the human–environment interaction more generally and the value of grounding theory building in real-world experience.

Format and Conduct of the Study

Using the case study of the whale rescue as our touchstone for critical comparison, then, we will successively explore each of the three theoretical frameworks: the dominant model, herein referred to as the "tradition of rationalism," and the alternatives posed by feminist theory's "care tradition" and by phenomenologist Martin Heidegger. The text is accordingly divided into four parts.

Parts I, II, and III, respectively, cover the three philosophical traditions; the short "whale stories" that introduce each of these sections highlight each tradition's defining characteristics, giving the reader a preview of the material to follow. The discussion of each theoretical framework in these three central sections includes two chapters: the first examines the philosophical tradition itself, focusing on the development of its environmental ethics "offspring" (the rationalist tradition and extensionism in Chapter 2, the care tradition and ecofeminism in Chapter 4, and the phenomenological tradition of Heidegger in Chapter 6), and the second is an interpretation of the whale rescue as rendered from the perspective of each tradition, recounted in the voice of an "imaginary ethicist" from within the tradition in question (Whale Rescue Story I in Chapter 3, Whale Rescue Story II in Chapter 5, and Whale Rescue Story III in Chapter 7).

Part IV consists of two summary chapters. Chapter 8 recaps what the study has revealed regarding the influence of the assumptions implicit in each tradition, the relative strengths and weaknesses of the explanatory power of each lens, and the contribution each rendition of the story makes to a more complete understanding of both the rescue event itself and the processes of ethical deliberation. And Chapter 9 considers the evolution of moral philosophy from dominant to alternative traditions and from inter-human to environmental theorizing in light of the current transition to the postmodern or ecological western worldview.

The presentation of the facts of the case study (Chapter 1), the three versions of the event as seen through the conceptual lenses (the "whale stories" found in Chapters 3, 5, and 7 and the excerpts that introduce of each of the three central sections of the text), and the quotes from and other references to the rescue scattered throughout the text are drawn from personal interviews with many of the participants in the rescue and from analysis of media coverage and other relevant documents. In order to keep the notes to a reasonable number, neither the direct quotes that dominate the various recountings of the event nor the frequent paraphrases that reflect common or conglomerate insights are individually cited. A list of the study participants and the reference documents can be found in the Appendix.

The research into the whale rescue was conducted during the summer of 1994. A central goal of the study involves seeking a better understanding of the operation of human moral agency; we are interested in how participants and observers defined the whale rescue as an ethical issue, how they perceived themselves and the whales, and how they made and then justified decisions. Therefore, I approached the case with the intent to develop "thick" descriptions of the event, descriptions that might in turn suggest patterns of meaning-making and serve as a foundation for theory building; and in conducting the research I assumed that different individuals would offer unique insights, in accordance with their personal experiences and perspectives. Given this set of assumptions and goals, a research approach largely grounded in the qualitative framework was the obvious choice. Exploring values, decision-making processes, and interpretations of personal experiences required not the elicitation of "Yes/No" responses to standardized survey questions but rather in-depth, open-ended, and highly personalized conversations or interviews, and constructing our three unique "stories" hinged on the sensitive interpretation of in-depth and thoughtful commentary. Qualitative research is descriptive and exploratory and focuses on understanding rather than on establishing facts or testing theory; it acknowledges that meaning is in large part constructed and variable across perspectives, and so it approaches data gathering and analysis with sensitivity to individual interpretations and to social contexts and in recognition of the interpretive role of the researcher as the primary research tool.[7] The "storytelling" orientation of the three "whale stories" (actually data-analysis chapters) reflects the creativity possible within the qualitative research framework and doubtless would not be found in an overly quantitative study.

Exploration of decision-making and values and sensitivity to the influence of conceptual lenses virtually require data collection via in-depth and open-ended personal interviews, a technique which yields highly personalized accounts and allows the researcher to "get inside" the interviewee's "head" and adopt his or her perspective. Attempting to be as comprehensive as possible, I interviewed (or at least had some level of access to the perspectives of) both participants and observers, workers on the ice and workers behind the scene, men and women, children and adults, natives and nonnatives, supporters and detractors, citizens of Alaska and of the Lower 48, and representatives of all parties to the "unusual alliance": the environmental community, the scientific community, the military, and the corporate sector (however, the virtual absence of the international perspective, especially that of the Russian participants, is an unfortunate shortcoming of the study). And given that this study took place almost six years after the rescue effort occurred, the personal interviews were supplemented with historical research; most of us have "20/20 hindsight," and even the most sincere recollections are inevitably influenced by the passage of time. Such uncertainties can be mitigated, to a certain extent, by "triangulation" among various data sources: thus contemporaneous accounts of the event serve as something of a check on recalled accounts and can be used in this way to maximize accuracy and minimize misinterpretation as a product of forgetfulness or after-the-fact reconstruction. By combining these two types of data-gathering techniques, we gain not only the benefit of distant reflection but also the accuracy and passion of the moment itself.

The case study is one of several approaches often used in qualitative research, and it is especially relevant to this study because it is geared toward the integration of theoretical concepts with concrete real-world data. The case study approach allows the researcher to "dig deeply" into discrete and relatively well defined events or situations. Although the researcher potentially sacrifices breadth of analysis or generalizability of conclusions using this method, he or she is able to accumulate data from a variety of sources and perspectives and to analyze the phenomenon in question in its real-world context, achieving significant (and otherwise unattainable) depth of understanding. The expectation is that even single instances of phenomena can tell us something important and can thus contribute to theory building, and the assumption is that human beings are sufficiently similar to allow the generation of general, if speculative, hypotheses from narrowly defined samples.

The one major ambiguity in the judgment as to the appropriateness of this particular case highlights the issue of generalizability as it is variously understood from the quantitative and qualitative perspectives. As "charismatic megavertebrates" (animals such as koala bears, pandas, eagles, and wild cats, which disproportionately capture both our imaginations and our charitable contributions) whales are thought to be atypical of environmental issues in general and thus might well be avoided as research subjects in the quantitative framework on the expectation that they would "screw up the data." From the perspective of the qualitative framework, however, this very atypicality is worthy of study in itself and is not an impediment to research since generalizability across the spectrum of the human–environment interaction is not the primary goal; additional insight is gained, in fact, by considering the extent to which the results are or are not transferable to the analysis of other issues (in other words, by distinguishing between significant and irrelevant differences as they affect the applicability of results beyond the case itself). In this instance, however, part of the goal of the study *is* to expand our thinking about environmental ethics issues in general, and it is important that the theory overviewed and developed herein is not perceived to be limited to any one species or any one type of human–nonhuman interaction. Therefore, although I deem the rescue event itself to be representative of environmental ethics decision-making in many significant ways and focus attention on it during the course of the following study, this emphasis does not exclude consideration of additional environmental ethics issues. Readers will thus find examples of a wide variety of other such issues scattered throughout the text, attesting to the broad-ranging relevance of the theoretical perspectives under discussion. And, of course, readers are strongly encouraged to use the study of this particular event as a springboard to further investigation of the full range of environmental ethics issues.

The realm of ethical decision-making and values is inherently elusive. Several interviewees echoed the sentiment that "It's difficult to sit here and figure out what motivated millions of Americans to think this is important"; and one of the participants in the rescue effort stated, "I could never quite really figure out whether we all kind of collectively lost our minds or if we actually did something good." The complexity and ambiguity that produce uncertainty (but also richness) in any study of decision-making may be further compounded in the realm of environmental ethics; in the words of one rescue observer: "People are deeply conflicted about their relationship with the natural world." And, of course, rarely is motivation singular and straightforward. This rescue, for example, undoubtedly could have been a public relations vehicle, whether for companies with an eye to their environmental image or for Inupiat whalers concerned about the growing public opposition to their whaling practices; but herein we will go to substantial lengths to uncover and demonstrate the various ways in which the effort was *also* (and perhaps primarily) a matter of ethical concern. The qualitative research paradigm is capable of (even committed to) viewing human behavior as the product of multifaceted and even internally inconsistent motivations and values.

Given such ambiguity in the subject matter, in the exploratory approach to data collection and analysis, and in the task of teasing out implicit assumptions and subtle implications, perhaps the best the qualitative researcher can do—and certainly what I have tried to do here—is to take an honest stab at making sense of the complexity and ambiguity without claiming to have found "the answer" or to have produced the definitive analysis.

This study is thus *an* analysis of the gray whale rescue, not *the* analysis; it is *an* assessment of the incompleteness of the dominant framework and the relevance of the alternative frameworks as potential supplements; it is *an* approach to the exploration of the influence of philosophical traditions as conceptual lenses.

It is an exercise in storytelling, which I trust resonates, at least in places, with your own personal experience of connection with significant people, animals, and places. It is offered in the spirit of respect for our own stories and those of others that Coles's work so admirably models and in the spirit of his conviction that in the sharing of such stories we find our own voices and we are all teachers and learners.

It's that feeling you get inside when you see a whale. You're excited, and you start smiling.

Chapter 1   The Case Study: The Grays of October and Operation Breakout

"What has happened here is nothing short of phenomenal" wrote one reporter from Barrow; "[it is] a story that has undeniable human interest." So what *is* it, this story of three whales? A story acknowledged as one of the top media events of its decade; a story of a three-week rescue effort with estimated costs exceeding $1 million; a story of cards and letters and donations of nickels and dimes sent in from children around the world; a story of Americans and Soviets working side by side; a story of technology and human muscle together battling Arctic ice; a story of people and whales and what united them in a place hostile to both for a remarkable moment in time.

To begin the story we need to take a mental journey to the last American frontier: the state of Alaska. Also known as "The Great Land," Alaska is home to more caribou than people, half of the world's glaciers (one of which is larger than the state of Rhode Island) and half of its active volcanoes, more than three million lakes (each larger than twenty acres), and North America's highest mountain (Mount McKinley, or Denali, "The Great One"). Its coastal waters host ten species of great whales, and its southern forests hold the vast majority of the U.S. bald eagle population. My own brief visit to conduct this research brought sightings of humpback whales, bald eagles, beluga whales, and puffins; roadside encounters with moose and dall sheep; the awe-inspiring sounds of a ponderously moving glacier; and even a meeting from a helicopter with a polar bear at the edge of the Arctic Ocean.

Alaska is a land in which the often uneasy relationship between humankind and nature is brought into sharp relief. The initial discovery of its fur-bearing wildlife by Russian explorers and hunters in 1741 was soon followed by the devastation of its seal and sea lion populations, and the potential for re-enacting this environmentally disastrous precedent remains to this day; where great natural bounty exists, there also lies the potential for great destruction in the name of "resource development," and Alaska perhaps symbolizes above all the importance of the fine line balancing these two sets of concerns. It is difficult even to think of Alaska without bringing to mind environmental controversies: the Exxon–Valdez oil spill of 1989, the perennial debate over opening the Arctic National Wildlife Refuge for oil and natural gas development, the wolf kills of the early 1990s, and the ongoing issue of subsistence hunting of endangered bowhead whales by native Eskimos on the Arctic coast. These and similar questions loom large among diverse peoples living in the midst of natural beauty and bounty and enacting in our own age the struggle, lost in too many other times and places, to balance the needs of humanity with the values of wildlife and wilderness.[1]

On our journey of the imagination we fly into the state's largest city, Anchorage, home to half of the state's population of approximately half a million. We wing our way through the Interior toward Fairbanks, a city with roots in the state's mining and gold rush history where researchers today explore wonders of a less tangible variety as they are revealed in the Northern Lights. And finally we head to the far north, across the oil and natural gas fields of the North Slope and into the village of Barrow, the northernmost U.S. settlement and the nation's largest Eskimo community. Sitting at the edge of the Arctic Ocean—which in this vicinity is ice-covered approximately nine months of the year—and with an annual precipitation of less than five inches, the Barrow area is an arctic desert; average temperatures range from twenty degrees (Fahrenheit) below zero in January (with wind chill reaching eighty below) to forty degrees in July, and for sixty-six days in winter the sun never rises above the horizon (the sun does not dip below the horizon for a similar period during summer, hence the name "Land of the Midnight Sun").

The people who call this remote land their own are the Inupiat (often referred to as Inuits). Their Eskimo ancestors settled the village around 2200 B.C. after crossing the Bering Land Bridge. They hunted marine mammals such as seals and whales from twenty-foot, seal-skin-covered, lightweight wooden-frame boats called *umiaks,* and they traveled inland on foot and by dog sled to hunt caribou, birds, and fish; clothing was made from animal skins, and their partially buried houses were constructed of bone, sod, and wood. Barrow has remained under continuous settlement by the Inupiat Eskimos for over four thousand years. Today its people struggle to balance the forces of modernization with the desire to keep alive the traditions and societal values that grew out of their historically subsistence lifestyle.

Perched along the coast just twelve miles southwest of Point Barrow, the northernmost geographic point of land in the United States, the village of Barrow served as a major supply point for commercial whaling ships in the mid-1800s. More recently, it has been the seat of government for the eighty-eight-thousand-square-mile North Slope Borough (the world's largest municipality, in geographic area) since its incorporation in 1972 following the passage of the Alaska Native Claims Settlement Act (ANCSA). ANCSA gave Alaska's native people control of over forty million acres of their traditionally claimed lands, including much of the North Slope, and as a result, taxation of the oil development on these lands is now the source of much of the Borough's income. Barrow's population of around approximately 3,500 is two-thirds Inupiat, and its blend of living tradition and modern services has made the village an increasingly popular tourist site. Barrow is perhaps best known in the Lower 48 as the site of the 1935 plane crash that took the lives of Will Rogers and Wiley Post, but among the people of the region it is an important center for conferences and other activities related to native rights issues.

Although it is an oversimplification to reduce the uniqueness and complexity of any people to one trait or activity, if there is a single defining characteristic of the Inupiat culture of Barrow it is the annual hunt of the bowhead whale (the *agviq* in Inupiaq); the village's symbol is a pair of bowhead rib bones rising from the shore and forming an arch at the edge of the Arctic Ocean, and in many ways the spring hunt is still the center of village life. Found only in the circumpolar waters of the Northern Hemisphere, bowhead whales (*Balaena mysticetus*) migrate northeast in the spring from the Bering Sea to the Chukchi and Beaufort Seas (to the west and east of Point Barrow, respectively) and then return in the fall, passing close to Alaska's Arctic Coast in open leads in the ice near shore during both phases of their journey. Although protected by the International Whaling Commission's moratorium on whaling, by the Marine Mammal Protection Act, and by the Endangered Species Act, the endangered bowheads are still hunted by the Inupiat people of Barrow and other Arctic Coast villages by special exemption for aboriginal hunters; in an attempt to balance the threat to the bowhead population with the "extinction" of Inupiat culture, of which the bowhead hunt is an integral part, a variable number of "strikes" are granted to the natives each year. A season with ten to fifteen whales "taken" is considered successful.

This now closely regulated practice carries on the substance and in many ways the form of subsistence hunting probably established by Eskimos of northwest Alaska before 1000 A.D. Although the traditional harpoon of wood, bone, and slate has been widely replaced by whaling guns and iron harpoons, whaling crews still camp on the ice for weeks awaiting the passage of the bowheads and then paddle out to confront them in the traditional *umiaks,* depending heavily on traditional knowledge of the whales and the environment for a successful hunt. The practice is surrounded by a set of rituals embodying the attitude of respect, which is believed to be the prerequisite for the whale's "giving itself" to the hunters. Killing a whale and successfully bringing the carcass to shore are communal efforts, and many villagers come out to help butcher the whale's body; the meat, blubber, bone, and baleen are shared among members of the community. One whale often feeds an entire village and all of its parts are used for food, handicrafts, or household items. Respect and leadership accrue to the man responsible for the kill. A successful hunt is a time of community-wide celebration and of respectful thanksgiving to the whales for giving themselves for the life of the village; a successful season culminates with the *Nalukataq,* or feast of celebration, which includes traditional dancing and the blanket toss.[2]

This discussion of the bowhead hunt as practiced by the Barrow Eskimos highlights several aspects of contemporary Inupiat society and thus helps to establish the context in which the whale rescue occurred: a time and a place marked by rapid and intense societal change that brings questions of cultural identity into sharp relief; by political and environmental controversy that increasingly draws the attention of an often belligerent outside world; and by an ancient, yet timeless, perspective on the greatest of marine mammals that encompasses, without internal incoherence, both brutal killing and respectful honoring. With this necessarily brief and general background in mind, then, let us turn our attention to the whale rescue itself, beginning with a brief introduction to the animals at the very center of the event: the gray whales.

The gray whale, *Eschrichtius robustus,* is one of about eighty species of cetaceans—whales, dolphins, and porpoises—still inhabiting our world's oceans. Like all cetaceans, gray whales are warmblooded, air-breathing mammals that give birth to live young. Now found only in the Pacific Ocean along the coast of North America, gray whales once inhabited the Atlantic Ocean as well, along both the European and the American coasts; a small remnant of the Korean population may still exist. Hunted nearly to extinction twice, once between 1850 and 1880 and again early in this century, the gray whale population gained full protection in 1946. Protection from commercial whaling has since been strengthened through the International Whaling Commission's moratorium and through the Marine Mammal Protection Act of 1972 and the Endangered Species Act of 1973. Under these regimes the gray whales have experienced a remarkable recovery, increasing at an average annual rate of 2.5 percent and currently numbering around twenty-two thousand (likely a historic high). During the summer of 1994, the gray whale became the first marine mammal to be "de-listed," to have its status downgraded from "endangered" to "threatened," in the history of the Endangered Species Act.

As their population has grown, the whales have expanded their feeding range further into the Arctic Ocean and in recent years have occasionally been found venturing into the unpredictably ice-filled waters of the Beaufort Sea to the east of Point Barrow. Unlike bowheads, grays are not adapted to ice conditions and are unlikely to survive if they remain in the polar waters as winter approaches; they lack the cartilage that bowheads have on the top of the head to break through thick ice from beneath, and since they generally need to surface about every five minutes to breathe (compared with the bowhead's thirty minutes) they cannot remain submerged long enough to traverse much distance beneath the ice.

Gray whales come to feed in the Arctic Ocean as the midway point in their annual migration; their round trip from Baja California to the Chukchi and Beaufort Seas and back is roughly a ten-thousand-mile journey, the longest migration route of any mammal in the world. Watchers from the shore regularly spot them traveling north from February to June and south again from October to February; the species is easily identified because they swim within a few miles of the coastline, because of their mottled gray color with lighter colored areas where barnacles attach to their skin, and because they lack the distinctive dorsal fins of most whale species. The whales feed on the rich floor of the Arctic Ocean during the summer months and calve in the warmer waters of southern California's lagoons during the winter; they eat very little if anything during the migration, surviving on fat reserves from the summer's feeding. (During the five-month feeding period they consume approximately sixty-seven tons of food and regain up to 30 percent of their thirty-five-ton average body weight.) As "baleen" rather than "toothed" whales, grays sieve vast amounts of water and sediment through keratin plates suspended in their mouths, filtering out the amphipods and other small bottom-dwelling animals on which the whales feed as they swim on their sides, parallel to the sea floor. Grays are not thought to form lasting social bonds; females mate with multiple males (usually between Thanksgiving and Christmas), and they and their young migrate in slower-moving groups separate from the males. With a gestation period of thirteen months, grays make their first migration even before they are born, *in utero.*

Although Russia regularly takes around 170 gray whales annually in accordance with the exemption for aboriginal peoples and periodically shares a small number of its allotted strikes with Alaska's coastal natives, grays are rarely hunted by the Inupiat Eskimos of Barrow. Sought as a food source only when the much-preferred bowheads are unavailable, gray whales are less desirable because of their heavily barnacled skin and relatively thin layer of blubber. They also fight more fiercely when harpooned than do the slower-moving bowheads and are thus more dangerous to hunt. The ferociousness of mother grays, in fact, earned the species the nickname "devilfish" from early whalers; females with calves are extremely protective and belligerent, and they have been known to attack whaling vessels that threaten the calves and to attempt to capsize research ships that approach too closely.

With this background on Barrow, its people and their traditional bowhead hunt, and the gray whale species in general, we are now prepared to take a look at the October 1988 whale rescue itself and to begin answering the question with which we began: "So what is the story of three gray whales?"

Barrow Discovers Three Gray Whales

"We were out there just trying to talk to them to see if they were all right."

"The first reaction was to blaze away with the camera, thinking we'd probably never see anything like this again and they certainly would go away soon."

It was with the practice of bowhead hunting that the rescue of the gray whales actually began. On Friday, October 7, 1988, Roy Ahmaogak of Barrow was exploring the ice-covered Beaufort Sea off the Plover Islands to the northeast of the village in search of signs of bowheads when he spotted whale spouts in the distance. Moving as close to the whales as he could on the thin ice, he discovered not bowheads but rather three young gray whales sharing two ten-by-twenty-foot holes in the ice about a hundred yards offshore, where the water depth was around forty-five feet. The whales tended to surface for periods of about three to four minutes, blowing around twenty times, between longer six-minute periods of sounding (diving and remaining below the surface).

Several biologists, whalers, and other Barrow townspeople visited the whale site over the next few days, but it had not yet occurred to anyone that the whales were actually trapped and would die unless they were able to escape soon: "There was no concern. Everyone assumed they would just swim out of here" and quickly find their way out to the open leads in the ice about five miles to the northwest and then through the leads to the open ocean approximately two hundred miles to the west.

On Wednesday, October 12, however, when the ice had thickened sufficiently to support their weight, North Slope Borough (NSB) biologists Geoff Carroll and Craig George approached the whales closely for the first time and began to think that the whales were indeed in trouble and probably were not going to escape unaided: "[that] was the first day we could really walk right up to the edge of the hole and get a close look at the whales. I think that's when it became real. . . . Everyone was somewhat horrified at their appearance."

Record cold temperatures and ice conditions had set in earlier and more quickly than usual that fall, forming landfast pack ice that had earlier trapped several ships in the area. During early winter as the pack ice forms, Arctic weather conditions change rapidly and it is not unusual for previously landfast ice to shift and thus create a "lead" of open water; it is through such leads, in fact, that the bowheads frequently travel in this region, and such an occurrence would have allowed these gray whales to escape easily, on their own. It is also true, however, that gray whales probably become trapped in the quickly forming ice every winter: the Inupiat often find a number of gray whale carcasses along the shore in the summer, and they assume that such whales die by becoming trapped in the ice, just as these three did, either because they fail to head south early enough or because the winter conditions set in too rapidly for them to react in time. As a Barrow resident mused: "I truly believe that had there been a satellite photograph of the area at that time there probably would have been fifteen similar scenes going on of whales trapped."

The World Meets the Whales

"The story of the whales, the ice and the people working to separate them has echoed around the world."

"The inundation of the media has begun."

Video footage of the whales was taken nearly one week after the initial discovery; Marie Adams, director of the NSB Public Information Office, interviewed George and Carroll on tape standing in front of the whales. They noted during the course of the brief interview that "It doesn't look very good for the whales," and they mentioned that harvesting the whales might be an option for the community. This original footage of the whales was sent to KTUU, an Anchorage television studio, where it was used in the nightly newscast on Wednesday, October 12.

On Thursday, October 13, the *Anchorage Daily News* ran a brief write-up on the whales; this very first article on the whales mentioned that the gray whale population had been increasing significantly for several years and that the whales' predicament could probably be explained in terms of their subsequently expanded range. Before the morning was over, calls from reporters around the world were coming in to the NSB Department of Wildlife Management nonstop.

Barrow hotels were soon overflowing as the media descended upon the village to cover the story of the whales. Local residents rented bed and floor space, as well as snowmobile rides out to the rescue site (for up to two hundred dollars per trip). This early coverage already described the whales' situation with words like "trapped" and "plight."

To Save the Whales

"The stranded gray whales are no longer alone in the effort to escape from their icy prison and find the open sea."

"That was just an innocent deal. The whales didn't have a chance unless somebody could break through the ice. We thought, 'Well, we got something that could break the ice.'"

After their close look on October 12 suggested that the whales were indeed trapped by the thickening landfast ice, NSB biologists called the Coast Guard to investigate their sending in an icebreaker to clear a path for the whales to the open leads; but neither of the Coast Guard's two icebreakers was available, one being in port in Seattle undergoing repairs and the other headed away from Alaska on a mission.

Rod Christ and Peter Leathard of Veco International, the oil field service company headquartered in Anchorage which manages much of the oil operation at Prudhoe Bay under contract to Arco Alaska (Atlantic Richfield), saw the story of the whales in the newspapers and realized that their ice-breaking hovercraft barge could be used at Barrow to break a path for the whales; Leathard called Ben Odom, senior vice president of Arco

Alaska, who, after a conference call to NSB Mayor George Ahmaogak, committed Arco to pay for fuel costs and to arrange for helicopters to tow the barge from Prudhoe Bay to Barrow. Leathard and Odom contacted the Alaska National Guard and requested the use of two CH-54 Skycrane helicopters; but only the Pentagon could authorize their use for this purpose since human life was not at stake. Odom contacted Alaska Senator Ted Stevens's office and obtained his support in seeking Pentagon approval, and Cindy Lowry, director of Greenpeace's Alaska office, contacted the Alaska Governor's office to request their assistance as well before herself flying to Barrow. The Pentagon authorized the use of the Skycrane helicopters as a training mission; the two Skycrane helicopters were flown to Prudhoe Bay, and crews prepared the hoverbarge, which had sat unused for over four years, for its journey to Barrow. Ron Morris of the National Marine Fisheries Service in Anchorage came to Barrow to coordinate rescue efforts and to deal with the media.

The Eskimo whaling captains in Barrow met to consider shooting and harvesting the whales as the humane alternative to leaving them trapped; they decided to allow the hovercraft rescue attempt as proposed by Arco and Veco and to support a rescue attempt with their own manpower. Because of cold temperatures and the constant movement of the ice, the two air holes had grown smaller, so whalers and biologists from Barrow took chainsaws to the site, enlarging the original holes and then cutting two new, larger holes nearby (allowing them to surface, not in the energy-intensive spyhopping action of raising their heads vertically out of the water far enough to expose their blowholes—usually done only infrequently in order to see above the surface—but rather in the normal motion of arching to the water's surface and exposing only their blowholes, not their entire heads).

Wildlife organizations and environmentalists around the country called the NSB with suggestions for freeing the whales, and as media coverage intensified, people all over the United States called their congressional representatives and Veco, predominantly expressing support for efforts to free the whales. Both the *Anchorage Daily News* and the *New York Times* noted that, although gray whales were officially listed as endangered, the survival of these three whales was insignificant in terms of the survival of the entire population, and *Anchorage Daily News* coverage began emphasizing the "unusual alliance" between the oil industry, environmentalists, the military, the government, and the Eskimos, who were all coming together in an effort to save the whales. One commentator summarized the spirit driving the effort and uniting the different groups: "The overwhelming opinion seems to be that we've got to try."

Getting to Know the Whales

"Daddy can we pet the whales?"

Capitalizing on the presence of the whales as a rare opportunity for scientific data collection, biologists Carroll and George videotaped the whales and set up underwater equipment to record their vocalizations; they analyzed blow intervals, duration of surfacing and sounding, respiration rates, frequency of vocalization, and the effects on these and other factors of darkness, stress, and the presence of helicopters.

The biologists named the whales in accordance with markings on their heads: Bone (the youngest, for the exposed bone on his rostrum [snout]), Bonnet (for a barnacle pattern on his head that resembled a cap), and Crossbeak (for a misalignment of the jaws that prevented him from closing his mouth completely). The Eskimos also gave them Inupiat names: Kannik (snowflake), Putu (hole), and Siku (ice), respectively.

The whale's heads and rostrums showed abrasions where contact with the edges of the ice had worn off skin and exposed underlying cartilage. Biologists estimated their length (Bone, 25–28 feet; Bonnet, 27–30 feet; Crossbeak, 30–35 feet) and determined that all three whales were quite young, one a yearling and the oldest at most three years old. The youngest whale was also the smallest and the one seemingly in the worst physical condition, with the most skin loss and abrasion.

Local school groups visited the site, and Barrow's third graders began class each morning by singing a verse they had composed:

Oh I saw three whales that were stuck in the ice
Arco said yes, and Greenpeace too
Veco's barge came out and smashed the ice
And the whales swam free!
Bump bah dah dump bah, gray whales!

Jim Nollman, an expert in interspecies communication, flew to Barrow and played music to the whales through underwater speakers. Several visitors stayed overnight with the whales on the ice, and almost all spoke to and touched them; for many, this was the first and only time they had had such close contact with a whale.

Operation Breakout

"We're trying guys. That's all I can tell you. We're really trying."

"The whale rescue operation is beginning to look like a military invasion."

Working in tandem, the Skycrane helicopters attempted to tow the hovercraft barge out of Prudhoe Bay (a dangerous undertaking for the pilots and crew and the first time two military helicopters had ever been used in tandem to tow a single object), but they made little progress and the plan to use the barge in Barrow was dropped.

The Eskimos decided that the best alternative would be to use chainsaws to cut additional holes in the ice and thus to move the whales gradually, hole-by-hole, toward the open leads. Several seven-by-thirty-foot holes were cut in approximately fifteen minutes each, but the whales were reluctant to leave their original two holes; tarps were placed over the original holes to darken them and they were allowed to freeze over so that the whales would be forced to move to the new holes. The first time they left their original holes, however, was said to be in response to Arnold Brower Jr. (whaler and Borough administrator) touching them and giving them verbal directions toward the new holes. The idea of moving whales through a series of holes cut in the ice by chainsaws was not new as this technique had been used periodically by the Inupiat hunters to tow a whale carcass back to shore; this was, however, the first time the technique had been used to move live whales. In the words of Arnold Brower Jr., "We've done that every other whaling season, maybe; it's the normal factor. But it's not a normal thing to coach live whales that way. That was some experience."

Omark Industries of Portland, Oregon and Standard Alaska Production Company sent chainsaws to Barrow, and Rick Skluzacek and Greg Ferrian of Kasco Marine flew to Barrow from Lakeland, Minnesota, at their own expense with six de-icers (water circulating devices) that were placed in the air holes to keep them from refreezing. The de-icers and lights were also used to attract the whales to each new hole. Bone played in the bubbles the machines produced, pushing his head in their midst and biting at them. The ice was one-and-a-half to two feet thick by this time.

Leaving the hovercraft behind at Prudhoe Bay, the Skycrane helicopters were flown to Barrow with a newly constructed fourteen-thousand-pound concrete "Ice Basher"; they began smashing holes in the ice, working inward from open water so as not to frighten the whales. A large ice ridge (miles long, a quarter mile wide, and grounded on the ocean floor such that the whales could not swim beneath it) began to be seen as an insurmountable obstacle in the path of the series of holes. In Florida, a net was constructed for the purpose of airlifting the whales out of the ice and carrying them over the pressure ridge, an alternative to be used only as a last resort.

On Friday, October 21, the three whales had reached the twenty-fourth air hole in the series when Bone, the youngest, disappeared beneath the ice never to surface again; he was presumed to have died, having succumbed to exhaustion and possibly pneumonia. This was a difficult psychological blow to many of the workers and a moment of sadness for hopeful watchers around the world; one reporter summarized, "When Bone, the smallest of the three, disappeared . . . the nation mourned." Also on this date, the *Washington Post* questioned whether the rescue effort had gone too far.

On October 22, a military C-5A Galaxy airplane (the largest plane ever to land on Barrow's airstrip) brought in Veco's 11.5 ton Archimedean Screw Tractor from Prudhoe Bay. Designed to cut through ice with screw-like pontoons, the Screw Tractor cut a fifteen-foot-wide channel between the pressure ridge and the line of air holes. Fifty-seven holes had been cut by this date, reaching one and a half miles from the original holes out toward open water, but the whales would not pass the twenty-fifth hole because the water had become too shallow (an underwater shoal reduced the water depth to a mere twelve feet). Inupiat elders advised the chainsaw crews to "start thinking like a whale" and to direct the series of holes along a line of deeper water; as the holes were rerouted the whales followed with notably increased enthusiasm and energy, surfacing in each new hole even before the cutting was completed. Following the new series of holes the two whales moved to within three miles of the ice ridge.

The ice was continually cluttered with radios, generators, lights, and wires, and by this time a hut had been set up near the site so that the rescue workers who were staying on the ice for long stretches and overnight would have somewhere to warm themselves and heat coffee. Families spent much time out on the ice, giving the site an air of festivity. NSB Search and Rescue helicopters ferried workers and reporters to the site. Polar bears were spotted in the area, and rescue workers began keeping guns close at hand in order to warn them off should they threaten the whales; volunteers stood guard over the whales around the clock.

The Soviet newspaper *Izvestia* announced on October 22 that Soviet icebreakers would be sent to assist with the rescue effort, in response to requests to the Soviet embassy from Greenpeace, the State Department, and the National Oceanic and Atmospheric Administration (NOAA); in the Soviet Union, television segments updated the progress of the rescue effort and the newspaper *Pravda* also covered the event. An orbiting NOAA satellite was activated early to provide assistance to the incoming icebreakers with analysis of ice conditions in the area (water and ice depth, fault lines in the pressure ridge, and distance to shifting open leads and ice-free ocean). Morris decided not to radio-tag the whales in order to track their progress should the rescue effort succeed because of the additional stress this would have inflicted on them (the tags would have to be shot into them) and because of the low probability of successfully tracking them very far (the antenna on the tags would likely snag on the ice and break off).

When the *Admiral Makarov* and the *Vladimir Arseniev,* both owned by the Far East Steamship Company, arrived a welcoming ceremony was held aboard the *Admiral Makarov.* Several reporters and rescue officials visited the Soviet ships (this was the first time an American aircraft had landed on a Soviet vessel) and learned that the story of the whales was a big news story in the Soviet Union as well as in the United States: "Our whole country is watching, just like everyone else," the Soviets reported. Captain Reshetov's first question upon arrival was reportedly, "How are the whales?" By nightfall, the *Arseniev* had cut a channel through the pressure ridge and across the remaining ice to within a quarter mile of the whales' position; ice chunks were left in its wake, however, and the path refroze quickly. News coverage of the unusual cooperation between Americans and Soviets quoted the Captain as saying, "It's very nice to work together to do some good between two people[s]," and labeled the rescue effort "an international cause celebre and a glasnost demonstration project." As the two Soviet icebreakers arrived, *New York Times* coverage quoted biologists around the country who were skeptical about the rescue effort; coverage also noted, however, that such reservations were not likely to slow the effort given the worldwide interest and support. *Izvestia* ran an article entitled "Last Chance for the Whales" detailing the course of the rescue operation.

With Tass, the official Soviet news agency, relaying frequent progress reports to the Soviet people, on October 26 an icebreaker flying both U.S. and Soviet flags cleared another two-and-a-half-mile channel from open water through the ice ridge and inland toward the last of the holes. The whales left the last hole in the afternoon, dove, and swam into the channel, and the icebreaker backed out in front of them. Over one hundred Barrow residents were on-site to witness and cheer the whales' escape, and helicopters carrying media crews to film the whales' escape hovered over the channel, frightening the whales into near panic.

Morris declared the whales freed and the operation a success once the whales entered the channel, but the chainsaw crews and biologists feared that the quickly refreezing channel would trap the whales again overnight. The NSB hosted a celebration party during the evening. At a press conference President Reagan responded to the announcement that the rescue effort had succeeded:

> I am gratified that the California gray whales have been released to the open sea. The human persistence and determination by so many individuals on behalf of these whales shows mankind's concern for the environment. It has been an inspiring endeavor. We thank and congratulate the crews of the two Soviet icebreakers who finally broke through to the whales. They were part of a remarkable team effort—by governments, individuals and business. Due to all these efforts, the whales have returned to the sea.

Chainsaw crews went out early on the morning of October 27 and found the whales in very poor condition, bleeding and gasping for air in a small hole in the refrozen icebreaker channel. The crews enlarged the hole and then cut a new series of holes out of the icebreaker channel, through the gap in the ice ridge, and toward the open lead about one mile out. The icebreakers cleared one final path, coming to within twenty feet of the whales' last hole; in the last hole, the whales were approximately three hundred yards from freedom.

Early in the morning of Friday, October 28, Alfred Brower, whaler and member of the chainsaw crews, went out to the site to check on the whales; he spotted them together near the channel, spoke to them, and then watched them dive in the direction of the nearby open lead. By most accounts he was the last person to see the whales. The two Soviet icebreakers headed home in the afternoon, and most of the media left Barrow on the evening flight. At a news conference NSB Mayor George Ahmaogak announced again that the whales seemed to be gone, and Morris, having returned from a search flight, reported spotting a whale track in the lead heading out toward open ocean: "We did see a whale track, an unmistakable whale track."

Several hundred individuals were ultimately involved in the effort, and toward the end there were an estimated 150 people actually on the ice with the whales. Over the course of thirteen days of approximately seven daylight hours, rescue workers (primarily Barrow natives) with chainsaws carved an estimated eight hundred tons of ice, cutting over one hundred breathing holes across a four-and-a-half-mile span of ice. Estimated costs of the rescue effort exceeded $1 million.

"Our Hearts and Prayers Are with You"

> "Biologists have been besieged with suggestions from people around the United States on how to get the whales to open water."

> "Not since little Jessica McClure slid down a Texas well had the world been so captivated by a struggle for life."

From Barrow classrooms to the White House, the entire country—and much of the rest of the world, in fact—was involved in the effort to rescue the trapped whales. President Ronald Reagan telephoned Colonel Tom Carroll, the Alaska National Guard official in charge of the military's involvement in the rescue, expressing his support of the rescue effort: "Our hearts and prayers are with you, and anything we can say or do to help you for the success of the operation, we'd be pleased to do it. We are very proud of all you are doing up there." White House News Summaries (synopses of the top five news events, which the President received twice daily) repeatedly included this story, occasionally as the top item. An estimated two hundred members of the press visited Barrow, some from as far away as Australia and Japan, and Dan Rather and other newscasters frequently noted as part of their daily coverage of the event that "the whole world [was] watching" the rescue effort unfold.

The public response to the rescue effort was nothing short of incredible. An Anchorage movie chain donated the proceeds of a showing of *Star Trek IV* (in which the Earth is threatened in the future because whales have become extinct), and a network news poll found viewers to be 90 percent in favor of spending tax money on the rescue if necessary. A Fairbanks T-shirt silk screen company printed over seven hundred whale rescue shirts and sold them out the first day. And a New York composer recorded a song, "Gentle Creatures," in honor of the whales and their rescuers.

The National Guard, the NSB, Greenpeace, Veco, Arco, NOAA—all the major participants—received countless telephone calls, telegrams, cards, and letters from people around the world in support of the effort; very few negative phone calls were received by anyone. In the words of Captain Mike Haller of the National Guard, "name a place and we probably got cards and letters from them." The local newspaper, the *Barrow Sun,* printed letters from (among other places) California and Canada, Ireland and Italy, Pennsylvania, Texas, Florida, Oklahoma, and Virginia. A letter from Switzerland came to the Borough with the Swiss equivalent of twenty-seven cents, and one from Italy came with three thousand lira. Over twenty-five thousand dollars were contributed to the NSB's Save the Stranded Whales Fund from individuals around the world (with another five thousand dollars contributed by Greenpeace), much of it in nickels and dimes that came enclosed with letters from school children (from science classes, a learning disabilities class, and many others). The following excerpts are representative of these letters:

> I am glad you are freezing your buns off for the whales. Thank you for caring so much. (from Joshua)
> We have heard about the whales on the news and have talked about the whales in class. (from Tony)
> I think whales are cool. (from Heath)
> I like whales. I hope you save them. I want my kids to see a California gray whale when I get some kids. (from Jason)
> I hope you all keep going and don't give up because the Gray whale is almost extinct and I want them to live. Thank you for doing this. (from Shain)
> I think it is very nice what you are doing. I couldn't imagine the world without our whales. (from Josie)

Many letters and phone calls also contained suggestions for freeing the whales:

> Why cant you use dimamit [dynamite] in spots? (from John)
> Maybe you could get a big army helicopter and strap them up and pull them out and take them to where there's no ice. . . . Please think about my suggestion. (from Benjamin)

And, finally, a poem entitled "Thank You" from Arthur in Wales (from which the concluding verse is excerpted) summed up the feelings of many people around the world:

> It's nice to know what man will do
> For creatures in distress
> To all who helped from one who cares
> My "Thank you" and "God Bless."

Into the Sunset?

> "[The media] said they were seen sort of trailing off into the sunset, but that's not the case."

> "I wish we had been able to capture that time [to see them actually swimming in the lead], but it's one of those things; you don't control Mother Nature."

At the conclusion of the three-week encounter with the gray whales, no one knew for certain whether they had indeed escaped into the open lead or made it to the open ocean almost two hundred miles to the west. Some of those who worked closely with the whales on the ice were optimistic that they had a good chance of escaping: they had demonstrated a strong will to survive throughout their three-week ordeal in the ice, and at the end they had only to swim under the ice for a few hundred yards in order to reach the lead. Further, National Weather Service satellite images indicated a series of ice-free leads almost all the way out to the Chukchi Sea, and local weather and ice conditions remained favorable for at least another week.

Others, however, worried that the stress of their ordeal and their poor physical condition (the skin damage and exposed cartilage and their undoubted exhaustion) boded for a less happy ending: the whales may have drowned right there at the edge of the lead, unable to find the open water, or they may have made it to the lead only to lack the strength and endurance necessary for the journey ahead of them (it would take at least five days for them to clear the icepack and reach the ocean). And, of course, the lead could have closed before they reached the ice-free ocean, trapping them once again in the icepack, or they could have encountered hungry polar bears as they navigated through the treacherous ice. If they made it to the ocean, they still faced the threat of predatory orcas (or "killer whales" that feed on grays, especially weak ones) and even Russian whaling ships during their journey south.

In an attempt to ascertain their fate, photographs of the whales were distributed to California tourboat operators, museums, and aquariums; the Marine Museum in Los Angeles trained one hundred volunteers to watch for the two whales and assigned them to whale-watching boats with pictures and lecture information for the tourists who were also asked to watch for them.

Two lone gray whales were spotted near Prince William Sound a few weeks after the conclusion of the rescue effort. A gray whale carcass washed on shore near Barrow during the following summer, but biologists George and Carroll examined it and determined that it was neither Bonnet nor Crossbeak. To date, a positive identification of the two whales (living or dead) has yet to be made, and the rescue effort's participants and observers remain divided on their opinions as to the whales' ultimate fate.

An Event to be Remembered . . . and Pondered

"A story not likely to be soon forgotten"

The Associated Press ranked the rescue as the top Alaska story that year (above Alaska's general economic downturn, problems with the salmon fishing industry, the search for seven lost walrus hunters in the Arctic, Anchorage's failed bid for the 1994 Winter Olympic Games, and the debate over opening the Arctic National Wildlife Refuge for oil exploration and development). And with coverage devoted to the whales in *Time, People, USA Today, U.S. News and World Report, Newsweek,* and *Life* as well as in *Audubon, Oceanus, The Economist, Spy, Macleans,* and in Alaska publications *Alaska Magazine, Alaska Airlines Magazine, Alascom Spectrum,* and *Uiniq: The Open Lead,* the whales were (in the words of the *Anchorage Daily News's* managing editor) "as well covered as any three mammals outside the royal family."

After the conclusion of the effort, President Reagan's office sent letters of congratulation and appreciation, such as the following, to the major participants.

> Please express my congratulations to everyone who has helped save the stranded California gray whales. This dramatic rescue effort has captured the world's attention these past October days—and perhaps reminded all of us of something essential about ourselves and our human nature.
>
> Thanks to this remarkable team effort by governments, individuals, the military, and business, these majestic creatures have resumed their journey south.
>
> Nancy joins me in sending heartfelt appreciation and again, congratulations to everyone on a job well done.

In the months following the rescue effort, the Cetacean Society International awarded "Cetacean Citations" to the major participants in the rescue effort, expressing appreciation for the effort on behalf of all the world's dolphins, porpoises, and whales. An Anchorage radio station sent a "thank you" telegram to the Soviet Union with the names of hundreds of listeners who called in wanting to participate. Jerry Faber, a chainsaw artist in Minnesota, sculpted a statue of the three whales out of a dead American elm tree and presented it to Vladivostok, the home port of the two Soviet icebreakers, along with letters of appreciation from citizens of the United States. In October 1989, Barrow hosted an anniversary celebration and awarded Certificates of Commendation to over one hundred residents for their participation in the rescue effort. And in October 1993, participants living in the Anchorage area held a five-year reunion, commemorating the whales, the rescue effort, and the friendships formed as a result of it by sharing their memories and memorabilia, their photographs, and their videotapes.

For many, the event will long be remembered for its personal and professional significance as well. The rescue effort brought Colonel Tom Carroll (the National Guard coordinator) and White House aide Bonnie Mersinger together; they fell in love during transcontinental telephone conversations that kept the White House updated and then married the following year; it won Fairbanks photojournalist Charles Mason an international World Press Photo award for his evocative photograph of parka-clad natives leaning over an air hole to touch one of the whales; and it launched the career of Tom McDowell, a videographer working with the National Guard whose performance during this event gained him the opportunity to cover the Exxon–Valdez oil spill, the children of Chernobyl, and a National Geographic expedition to Antarctica with Jacques Cousteau. At least two other books have been published on the rescue: a children's story written by Giles Whittell and illustrated by Patrick Benson entitled *The Story of Three Whales* (Milwaukee: Gareth Stevens Children's Books, 1989) and a book by journalist Tom Rose suggestively titled *Freeing the Whales: How the Media Created the World's Greatest Non-Event* (New York: Carol Publishing Group, 1989).

And how did the various participants and observers make sense of this remarkable event? As Rose's subtitle indicates, for some it was a "non-event" driven solely by the media and by the public's sentimental interest in animals and in drama; for some, it was a "fiasco," albeit one that "started with the noblest of intentions"; for others it was "an absurd waste of money by any standard"; and for still others it was "the single most irritating experience" of their careers. Many questioned whether public relations and profit rather than humanitarian concern were the driving motivations for many of the participants, and many wondered if the effort was not more about proving our ability to "win" the struggle against nature than about the whales themselves. Representing the alternative, and probably predominant, opinion, however, the *Chicago Tribune* asked, "Were the million-dollar whales of October merely an aberration, a TV show, as some are suggesting? Or has something deeper happened? Has society finally decided to stop the killing?" President Reagan suggested that the rescue effort "highlighted mankind's concern for the environment," and the majority of observers seemed to be proud of their fellow human beings for showing compassion and responding to the whales' plight. A letter to the *Los Angeles Times* expressed this position: "In a world where man's inhumanity towards animals appears to be the rule, it was a welcome exception."

Whatever meaning or significance the story of the whale rescue may or may not have had in the minds of participants, observers, and commentators, it was without doubt "a story not likely to be soon forgotten." In light of the central role to be played by the case in this study, it probably goes without saying that I disagree with the opinion of one observer that "there is just no substance to this story anywhere." Given that it has been called "the most extraordinary animal rescue effort ever undertaken" and that "it was one of the top human interest stories of the century," I take the position that it was not a meaningless occurrence. I agree with one of the on-site observers that "it really did say something about people and about what it means to be human, and it is important to try and figure out what that is."

And that is one of this text's three primary purposes: to "figure out" why this remarkable rescue effort unfolded as it did and what it has to tell us about ourselves. Now that we have the basic facts of *how* the story unfolded, throughout the remainder of the text we will explore *why* it happened as it did, from the theoretical perspectives offered by our three "lenses" in conjunction with the "real-life" perspectives of the people who were actually involved. What, then, are the central questions to be asked of this case, the questions that will help us come to a clearer understanding of the rescue effort as an environmental ethics issue? While each of our three frameworks will take a different approach in conceptualizing this as an environmental ethics issue (these differences will be explored in depth in the chapters to follow), there are certain core questions that each framework, if it is to be comprehensive in its interpretation of the case, will have to answer. Among these are the following four:

1. Why was the rescue effort the appropriate course of action?
2. What was the significance of their being *whales* rather than another type of animal?
3. What was the basis for opposition to the rescue effort?
4. What are the implications of this rescue effort for environmental issues in general?

Chapters 3, 5, and 7 each will offer a composite answer to each of these questions—a composite drawn from the experiences of participants in, observers of, and commentators on the rescue effort—from the perspective afforded by our three conceptual lenses or frameworks. The "imaginary ethicists" speaking from within each framework will explain the event differently and will consider more specific questions unique to the individual lens, but each will do so in the context of these same four central questions, so that we will have a constant basis for comparing and contrasting the three frameworks.

Before turning to the first of our frameworks, however, it is fair to question the relevance of the philosophical perspective to this rescue event as a particular manifestation of the human-environment problematique. One of the participants in the rescue effort expressed just such a challenge to couching the rescue in the terms of environmental ethics decision-making: "It is not clear to me whether there are any real ethical considerations. The whales were there, they were trapped, and a decision was made to try and get them to open water. I don't think that's an ethical question." Similarly, in many people's minds the whale rescue is a story of technology or is a political or a media event, without necessary reference to the field of philosophy or its subdiscipline of ethics. Interdisciplinary environmental studies, however, rest on the premise that environmental events are neither simply technological, political, scientific, or economic in nature (although they are, of course, each of these) nor the sum of these perspectives; insofar as they confront us with difficult decisions in which we must make trade-offs among different values and then justify these decisions in the realm of public dialogue, they are also fundamentally philosophical in nature. Echoing this claim, Joseph DesJardins, in his introductory environmental ethics textbook, writes:

> Environmental and ecological controversies raise fundamental questions about what we as human beings value, about the kind of beings we are, about the kind of lives we should live, about our place in nature, and about the kind of world in which we might flourish. In short, environmental problems raise fundamental questions of ethics and philosophy.[3]

It should be clear from the events of the whale rescue that it does indeed raise these types of questions. While the rescue can certainly be viewed from other disciplinary perspectives, as a media or a public relations event, for example, it is also—and I claim, fundamentally—a story about what we value in the natural world, about the character traits and kinds of behavior we deem praiseworthy, about how we make decisions regarding our role in nature, and about how we variously conceive of ourselves as "apart from" or "a part of" the natural environment. It raises difficult questions about justifying our intervention (and nonintervention) in the "natural" course of life and death, about special obligations we may have incurred due to our history of overexploiting certain animals (such as we have with many species of whales), about our tendency to value some species to the exclusion of others; and about how we choose to allocate our resources of time, effort, and money across the spectrum of our social and environmental concerns. Thus, the case study of this whale rescue is above all an appropriate candidate for analysis from the perspective of moral philosophy.

Out of immediate danger. Revealing abrasions caused by breathing through small airholes in the ice before being discovered, a gray whale surfaces comfortably in a larger airhole cut by Barrow residents. Such intriguing and moving images caught the attention of the world and launched a three-week, million-dollar, international rescue effort.

A quiet moment on the ice. The three young whales, Bone, Bonnet, and Crossbeak, frequently drew near as people approached the airholes. For many of the participants in the rescue effort, such face-to-face encounter fostered empathy and a sense of relationship that strengthened commitment to the rescue effort.

Jens Brower, an Inupiat child and member of a local whaling family, observes the rescue operation. School groups from Barrow visited the whales at the rescue site, and children across the country sent drawings, letters, and donations to the rescue workers.

A whale surfaces beneath the Arctic moon in the vast expanse of landfast ice.

Part I   Philosophical Tradition I

When I remember those trapped whales, I always think about the Jessica McClure story too. The two events were so similar: big rescue efforts, lots of media attention, public outcries. And together they show how far we have come, ethically speaking. I mean fifty years ago who would have thought that rescuing three whales would one day be seen in the same light as rescuing a little girl?

But I guess whales are people too. They are intelligent and they may even have emotions and underwater societies; they live in family groups, and help one another when need be, and play in the moonlight and sing complex songs to one another. Whales are "in many ways so much like man," and it is only prejudice for our own species that makes us think "we are the most important living organisms on earth, that our problems and our welfare are somehow more important than any other creatures." "Every whale is worth the effort" of saving just as every child is, and it is a testament to our ethical progress that "there are some people who see us all, baby girls and whales, as these sentient beings that all matter."

And really, once you get rid of all that mushy, sentimental stuff—you know, the close-up pictures of the McClure family crying and the whales gasping for air, the media's pulling on our heart strings to sell newspapers— once you think about it logically, both cases were pretty simple and clear- cut. The same basic principle was at stake: "there's a respect for life hope- fully all people would have." A life is a life, and we have a duty to save life if we can, whether it's a whale or a little girl. "Knowing that other creatures have an inherent worth just like we do, we have this moral obligation to help out our fellow species." "The whales were suffering and we felt a duty to do what we could to alleviate that suffering."

And we had an extra obligation to the whales as well: "we humans owed those whales the chance for survival since after all it was we who caused them and other forms of wildlife to be placed on the endangered species list." Since it was we who inflicted such injustice on them in the past, it does seem that we have the duty to "balance the equation a little bit."

I really didn't understand, though, why the Eskimos or the Soviets judged themselves to have such an obligation, given the obvious disregard for the inherent worth of whales reflected in their hunting practices. "It does not seem to make a whole lot of sense that they're out there hunting whales and here they were saving three of them." It's inconsistent.

It's ironic that the Eskimos and the Soviets participated in the effort while so many other people opposed it. Remember those editorial cartoons of homeless people trying to get help by wearing whale costumes and that one showing people dying of starvation while so much attention was being focused on the whales? That's the hard part about trying to do the right thing: there are always winners and losers—my right to smoke versus your right to clean air, the fetus's right to life versus the woman's right to choose, dying whales versus dying people—and it is difficult to prioritize when we have to choose one over the other.

Given that we have just as much of a duty to the whales as to one another, though, for me the tough part of the decision was not whether the money should have been spent on people instead but rather whether we should be spending all that money on just *three* whales, and whales "that were not even becoming extinct" at that. "These weren't important animals in the larger scheme of things: they weren't among the last of their breed and at no point did they make any difference in the gene pool." "There were certainly far greater threats to the whale population than these three being caught in the ice," and a million dollars could have helped address these threats and save a whole lot more whales: educating people, working to outlaw all whaling, helping to set up that whale sanctuary in Antarctica. The thing is, you could say the same about Baby Jessica: that money could have gone a long way toward immunizing children or toward a hospital pediatrics ward. What's more important, then, the individual or the larger whole? Rescuing three whales or putting that money to work on behalf of populations or ecosystems or diversity itself?

I don't know the answer there, but it sure did seem that very few people were thinking that way; in fact, very few people seemed to be thinking at all. Everyone was just caught up in the emotional appeal of the story. They even considered shooting polar bears to protect their "pet" whales, even though reason clearly indicates that "you don't shoot one endangered species to save another." And, unfortunately, the environmentalists who were promoting the rescue did little to counter the "Oh, aren't they cute" and "Isn't that pitiful" responses to the whales. "Cashing in on sentiment to save these three whales, however, was not sensible." "What is needed is not sentiment but a much deeper respect and reverence for life." "For the long haul, environmentalists must respect rationality and hope to inform the public."

I think what it all came down to for me was this: that whale rescue showed that we just may be realizing, finally, that "we are just another link in the chain." We humans aren't the only creatures on this planet that deserve moral consideration and we can't just do whatever we want without taking the rights and interests of other creatures, whether individuals or the larger systems they comprise, into account. It wasn't so long ago, after all, that we thought we could treat women and minorities any way we pleased, as if they were somehow less important. Not too long ago, in fact, we might easily have left a baby girl stuck in a well to her fate. Wouldn't it be interesting if, a hundred years from now, saving whales had become as uncontroversial as saving little girls had by the 1980s? We certainly seem to be headed in that direction.

The sole aim . . . is to seek out and establish *the supreme principle of morality*. . . . Do we not think it a matter of the utmost necessity to work out for once a pure moral philosophy completely cleansed of everything that can only be empirical . . . , [to] distinguish motives which, as such, are conceived completely *a priori* by reason alone and are genuinely moral . . . ? . . . Morals themselves remain exposed to corruption of all sorts as long as this guiding thread is lacking, this ultimate norm for correct moral judgement.[1]

The fact is, that each time there is a movement to confer rights onto some new "entity," the proposal is bound to sound odd or frightening or laughable. This is partly because until the rightless thing receives it rights, we cannot see it as anything but a *thing* for the use of "us"—those who are holding rights at the time. . . . There will be resistance to giving the thing "rights" until it can be seen and valued for itself; yet, it is hard to see it and value it for itself until we can bring ourselves to give it "rights"—which is almost inevitably going to sound inconceivable.[2]

Chapter 2   The Tradition of Rationalism

We begin with the exploration of the first of our three lenses: the dominant framework in (western) moral philosophy, referred to here as the *tradition of rationalism*. Rather than attempt to peruse the entire canon, which would doubtless fill many volumes and is not to our purpose in any case, we will restrict our discussion of this framework to one concept which, wide-ranging in both its applications and its implications, embodies the core characteristics of this perspective and thus allows us to examine it in a more manageable microcosm. This representative concept is referred to as the *moral point of view* (MPV), and our examination of it will trace its evolution, and simultaneously that of the tradition of rationalism in moral philosophy itself, from the early Greek philosophers through its twentieth-century incarnations in both mainstream and environmental ethics. Although our focus here will be on the tradition of rationalism as it is embodied in environmental ethics theorizing, we will begin with a brief overview of the history of traditional, interhuman, ethical theory that gave rise to and in many ways defined the structure of this environmentally oriented "offspring."

The Moral Point of View

Many contemporary ethics texts begin with an injunction similar to the following: "Ethics requires us to go beyond 'I' and 'you' to the universal law, the universalizable judgment, the standpoint of the impartial spectator or ideal observer."[3] The concept of the MPV plays a central role in mainstream ethical theory, a role that is often compared with that played by the scientific method in the natural and physical sciences. Just as the scientific method offers the investigator a procedure for defining and answering scientific questions, so too does the MPV assist the moral agent, the person faced with an ethical dilemma, in asking and answering moral questions.

The MPV is, most simply, that perspective which we must consciously adopt during the process of ethical deliberation in order to ensure the moral legitimacy of the conclusions we reach. The scientific method assures scientific legitimacy by advocating a stance of complete objectivity and a procedure based on principles of logic; it also eliminates from the investigator's consideration such factors as the agenda of the agency funding the research, the purposes to which the research results will be put, and the prospects for personal gain that may hinge on the outcome. In a similar fashion, the moral point of view consists of elements to be avoided and others to be achieved. The moral agent's personal preferences, feelings, and uninformed opinions are not to be taken as grounds for moral judgments any more than are the opinions of the majority or the commands of a moral authority (humorously but not inaccurately defined in my environmental ethics classes as "God, the law, or Mom"); the ideal judgment will be the product of complete knowledge of the facts and clarity in language and in reasoning processes and will be based on the impartial, emotionally "cool" application of valid, universal principles. The MPV can be thought of, in simple terms, as the bird's eye view of an ethical dilemma in which the bird knows all, cares for all or none equally, disregards the opinions of other birds, and legislates, in a deliberate and consistent manner, the highest principles of birddom. The MPV both operates out of and reinforces the autonomy of the rational agent; as decision-makers we stand alone and independent at this vantage point, our reasoning abilities rendering us fully capable of reaching and then justifying valid ethical conclusions. It is simultaneously a perspective and a procedure, and without it our ethical deliberation is subject to a whole host of illegitimate influences that undermine the validity of our judgments.

As we will see in this chapter's necessarily brief and simplified overview of the evolution of the MPV concept, it is these very elements—objectivity, individual autonomy, the application of abstract principles, the primacy of reason over emotion, and an emphasis on universal principles, among others—that characterize the perspective on ethics offered by the tradition of rationalism.

The Emergence of Rationalism among the Early Greeks

The roots of this tradition are to be found in the works of those early Greek philosophers to whom we trace the very foundation of western civilization, especially that of Plato, "whose thought would become the single most important foundation for the evolution of the Western mind."[4] At least three different strands of thought, which stand out in sharp relief against the backdrop of a Greek world otherwise subject to supernatural explanation and peopled with capricious mythical figures, combined to lay the foundations of rationalism upon which Plato built.

1. *The assumption of an order in nature which the human mind could comprehend.* The first element of rationalism emerged in the seventh-century B.C. philosophy of Thales, Anaximander, and Anaximenes. Perhaps the earliest scientists, these men shared an assumption about the world around them that profoundly influenced the subsequent course of scientific investigation and that formed the cornerstone of the tradition of rationalism: they sought to look beneath the apparent movement and change characterizing the physical world for an assumed underlying orderliness or rationale in creation. (Thales, for example, hypothesized water as the principle unifying the world's movement and substance; and Anaximenes posited air as the primordial matter and explained how its thickening and thinning produced heat and cold and the different substances.) Two significant elements were thus introduced: not only was the world itself seen to possess some sort of rational order, some unifying principle in terms of which it all "made sense," but the human mind was understood to be capable of discovering this principle through the use of reason.

2. *The privileging of abstract reason over sensory experience.* The second strand of thought was introduced by Parmenides a century later. Struggling with the questions of what is real and how to explain change in light of the assumption that reality was intelligible, Parmenides discounted reliance on the "evidence" of our senses as an avenue to understanding the true nature of reality and instead insisted that reason alone could apprehend reality. Noting the apparent tendency of substances in the world to come into being

and pass away and insisting on the logical impossibility of a thing that *is* coming from or becoming a thing that *is not* (since "nothing" cannot exist), he concluded that such sensory experience of change was illusory. Observation yielded false "opinion" about the nature of reality, whereas reason alone could transcend such illusion and apprehend "truth"; therefore, investigation must be conducted in the realm of abstract logic unhindered by apparently contradictory empirical findings. Reason and experience were thus cast as opposites and Parmenides chose to rely on the former, setting for early philosophy the problem of how we know what we know and how we legitimate claims to knowledge. Parmenides thus introduced dualism of the intellect and the senses as avenues of knowledge, skepticism of the merely empirical, and reliance on abstract reason and logic as a second cornerstone of the incipient tradition of rationalism.

3. *The application of reason to questions of morality.* Socrates took the emerging tradition of rationalism and bent it toward a new purpose: not toward the study of the physical world but toward investigation into how the human being should live. More important than unveiling the mysteries of material phenomena, reason illuminated the very essence of the "good," and it was only by discovering the nature of the good that one could truly lead "the good life." Morality was thus taken to have a rational basis in that one could discover, through the use of reason, that which was good; reason would then also lead one to do the good since it was only through misunderstanding the nature of the good (via an error in reasoning) that one would commit evil. For Socrates, "A truly happy life is a life of right action directed according to reason. The key to human happiness, therefore, is the development of a rational moral character."[5] As a teacher, Socrates focused on the development of a method for reasoning one's way through such questions. The *Socratic method* is this process of critically thinking about ethical propositions, usually taking the form of an argumentative dialogue between two people; it involves asking a series of questions and critically analyzing the implications and assumptions associated with the answers to highlight inconsistencies in the reasoning process and thus to arrive at the truth.

Tarnas sums up the significance of these strands of thought in laying the foundation for Platonic rationalism: "By implication of these early philosophical forays, not only the gods but the immediate evidence of one's own senses might be an illusion, and the human mind alone must be relied upon to discover rationally what is real."[6] Reason, abstracted from experience, was the most useful tool human beings possessed in our quest to understand the world, which itself possessed a rational and discernible order. And the use of reason would clarify how we should live; indeed it was the carefully logical examination of such questions that gave meaning to human life, as suggested by the Socratic dictum: "The unexamined life is not worth living."

Platonic Rationalism

The tradition of rationalism was thus well begun by the time Plato began studying under Socrates around 405 B.C. It had initiated an important shift in the tenor of Greek thought, away from supernatural explanations of a largely random world and toward the investigation of an orderly world that awaited only the attention of human reason for its sensible unveiling; and it had fostered the development and refinement of methods for increasingly rigorous, logical reasoning. It fell to Plato, however, to draw these elements into a coherent framework and thus to complete and institutionalize the foundation for so much of western civilization, including the tradition of rationalism within moral philosophy.

The central tenet of Plato's thought is that of the *Forms*. Building on Socrates' belief in an abstract, universal concept of "goodness," which transcended particular manifestations and thus served as an objective foundation for ethical deliberation, Plato posited the existence of the Forms as a collection of several such absolute Ideas that transcended the material world while giving it meaning. Thus an object in the world that is a circle, or a horse, or red, or beautiful is actually a concrete expression of the perfect Form of Circle, Horse, Red, or Beauty. Objects in the physical world derive their particular being from their "participation in" the universal Forms. Although the objects partaking in the essence of the Forms come and go, changing with time, the Forms themselves are immaterial, absolute, and changeless—and thus more fundamental, more real, truer. Reality is thus fundamentally dualistic in nature, with the realm of the Forms standing above and beyond (and essentially separate from) the material world.

Empirical observation, of course, reveals only physical objects and their particular properties; it is the purview of the intellect, the rational faculty that can conceive of such a high level of abstraction, to ascertain the essential reality that lies behind and gives meaning to material appearances. Thus the human being is also understood in dualistic terms: there is Reason and there are Appetites, nonrational physical desires, and it is Reason that should control and legislate our behavior. It is only through the rigorous training of the intellect, in fact, that the philosopher can attain awareness of the Forms and thus see through deceptive appearance to transcendent truth.

Plato's metaphor for this change in perspective is presented in the *Allegory of the Cave*. Most of us live our lives as if chained to a wall inside a cave, observing shadows that are cast on the opposite wall by movement around a fire outside; not knowing of the world beyond the cave, we take the shadows to be reality. A man who escapes from the cave and ventures into the world beyond will come to understand the shadows for what they really are, mere reflections of the true reality that is beyond the reach of those still trapped in the cave. This is the man in whom reason has reached its fulfillment: the philosopher. Integrating dualism, universality, and rationality, this doctrine of the Forms is one of many instances in which Plato drew together and further refined many of the elements of early Greek rationalism.

Although he did not use the phrase "moral point of view," Plato posited perhaps the earliest version of this concept in the *Crito*.[7] In this dialogue Plato chronicles a conversation between his mentor, Socrates, and Crito, a friend who has come to the Athens prison in the hope of convincing Socrates, convicted of "impious" teachings that "corrupted the youth" and awaiting the death penalty, to cooperate in an escape attempt. The usefulness of the dialogue for our present purposes lies in its detailed account of the reasoning process Socrates leads Crito through, step by step, in reaching and justifying his decision to remain in prison. Socrates' deliberation hinges on three major elements.

1. *The importance of making a carefully reasoned decision rather than an emotionally driven one.* Socrates refuses to be swayed by Crito's concern regarding the uncertain fate the philosopher's children will experience should he indeed be killed, claiming, "We must consider whether or not it is right for me to try to get away [and] . . . not think more of . . . children or of . . . life or of anything else than . . . of what is right."

2. *The need to rely on one's own principled reasoning rather than on the opinions of others.* Crito uses public opinion as an argument in favor of Socrates' escape, claiming that "one has to think of popular opinion [and] . . . it will look as though . . . we have let you slip out of our hands through some lack of courage or enterprise on our part, because we didn't save you, and you didn't save yourself, when it would have been quite possible and practicable, if we had been any use at all." Socrates' response, "Why should we pay so much attention to what 'most people' think? . . . What we ought to consider is not so much what people in general will say about us but how we stand with . . . right and wrong, . . . the actual truth," reveals the philosopher's disdain for allowing either the judgments of others or fear of the public's condemnation to replace one's own deliberation process.

3. *The application of general principles to the particular situation (deduction).* In reasoning through the question of whether it is morally wrong to attempt escape, Socrates appeals to three abstract rules whose validity as moral principles he also defends: harming others is wrong, violating one's agreements is wrong, and disobeying those in the role of parents and teachers is wrong. The argument then proceeds by Socrates demonstrating how his escaping from prison would be a particular instance of harming others (by showing disrespect for the laws of the state), breaking promises (the promise to obey the state's laws made implicitly by a citizen who chooses to live within its borders), and disobeying parents/teachers (a role filled by the state through its regulation of marriage, childraising, and education). Socrates thus resolves his dilemma, concluding that it is wrong to attempt escape, by applying general principles of right and wrong to his particular situation.[8]

Plato's discussion of appropriate ethical deliberation, then, emphasizes the necessity for deliberate, autonomous reflection on the part of the moral agent, specifically reflection that is not swayed by changing personal circumstance, that is deductive and free from the influence of emotion or of uninformed public opinion, that is driven by an overriding concern for maintaining the individual's moral integrity, and that is sufficiently sound to stand the test of logical scrutiny. These elements are at the very heart of the moral point of view as it has evolved since the time of Plato, as is the core commitment that moral judgments not be made in an arbitrary manner and the conviction that they gain legitimacy insofar as they adhere to certain standards of rationality. As we will see in the following section, modern moral philosophy has taken up these same assumptions and has continued to refine the concept of the moral point of view.

Rationalism in Modern Moral Philosophy

Cartesian Rationalism

In many ways it was René Descartes who, in the first half of the seventeenth century, ushered the legacy of Platonic rationalism into the realm of modern philosophy. Drastic revisions in the western worldview had been brought about in the intervening centuries by Christianity's displacement of paganism; by the Age of Exploration; by Renaissance advances in medicine, art, and literacy; and by the Copernican revolution's radical challenge to the geocentric model of the universe. In part through the rediscovery of many of the classic texts that had been preserved in Islam throughout the fall of

Rome and the subsequent Dark Ages in western Europe, many elements of the rationalism of early Greece persisted as defining elements of the "modern" worldview: elements such as the quest for absolutes or for universal principles, the primacy of reason over both empirical observation and subjective experience, and the conceptualization of reality in dualistic terms.

It was the very uncertainty of the sixteenth and seventeenth centuries, in fact, that in many ways prompted the development of Cartesian rationalism and thus facilitated the transition of these concepts from ancient to modern philosophical theorizing. Frustrated by pervasive confusions and contradictions, Descartes took as his goal the establishment of a rational basis for certainty, not only in science but also in morality. He imported from his mathematical training the abstract procedure of basing increasingly complex formulations on the simplest of fundamental axioms and thus developed for philosophy a method of determining absolute truth, a method that was not subject to the illusory and contradictory "evidence" of the senses. The search for truth, whatever the realm of investigation, should always be accompanied by skepticism—the seeker doubting any claim to truth that was not self-evident—and should proceed by deduction from this small set of irreducible and absolute first principles.

The one fact Descartes trusted as absolutely certain was that he was engaged in thought, that he was aware of his own doubting, and thus that he did indeed exist: hence his fundamental axiom, *Cogito, ergo sum,* "I think, therefore I am." This essential principle, held against the backdrop of Platonic dualism, led Descartes to a conceptualization of the world that has come to dominate modern thinking and that thus pervades the tradition of rationalism in moral philosophy: the process of thinking by which we are assured of our own meaningful existence also defines us as fundamentally distinct from the nonrational material world. As Tarnas summarizes Descartes' core insight,

> The *cogito* also revealed an essential hierarchy and division in the world. Rational man knows his own awareness to be certain, and entirely distinct from the external world of material substance, which is epistemologically less certain and perceptible only as object. Thus, *res cogitans*—thinking substance, subjective experience, spirit, consciousness, that which man perceives as within—was understood as fundamentally different and separate from *res extensa*—extended substance, the objective world, matter, the physical body, plants and animals, stones and stars, the entire physical universe, everything that man perceives as outside his mind.[9]

Thus Descartes built on the ancients' understanding of the superiority of the rational mind by explicitly adding hierarchy to dualism: privileging abstract reasoning over empirical observation, facts over values, and humankind over nature. Everything beyond the human mind was rendered an "other," an object, generally defined in mechanistic terms and susceptible to measurement and manipulation. Justly fulfilling the destiny first embarked upon by the ancient Greeks who had confidently set the task of unveiling the workings of natural and physical phenomena, "man" had only to further develop his intellect and his will to fully realize his role as "master and possessor of nature."

The implications of Cartesian rationalism for modern moral philosophy, specifically as embodied in the MPV, will be explored below in two central bodies of modern ethical theory: utilitarianism and Kantian ethics. Each of these perspectives on moral philosophy embodies, in its own unique way, the core assumptions characteristic of the tradition of rationalism, and thus each validates an understanding of the experience of moral agency in terms of reason, dualism, abstraction, and individual autonomy.

Utilitarian Ethical Theory

Aware that the multiplicity of religious perspectives introduced by the Reformation was fragmenting and potentially undermining ethical discourse in eighteenth-century England by offering disparate perspectives on truth and morality, philosopher Jeremy Bentham established a new body of ethical theory that he hoped would be free of such divisiveness. In the spirit of Cartesian rationalism, according to which the search for truth was to be grounded in the simplest and most self-evident of first principles, Bentham based his new theory, *utilitarianism*, on the one aspect of the human experience he deemed truly universal: we experience both pleasure and pain and are attracted to the former and repulsed by the latter. "Two sovereign masters," he wrote in his central treatise, *Principles of Morals and Legislation*, "pain and pleasure, . . . govern us in all we do, in all we say, in all we think."[10] Pleasure, which Bentham and his follower John Stuart Mill equated with happiness, was the one thing we could all agree on as intrinsically "good" and thus could be the only possible universal standard for judging right and wrong.

Utilitarianism, in Bentham's words, is "that principle which approves or disapproves of every action whatsoever according to the tendency which it appears to have to augment or diminish the happiness of the party whose interest is in question."[11] Put a bit more simply by Mill in *Utilitarianism*, it is "the creed which . . . holds that actions are right in proportion as they tend to promote happiness, wrong as they tend to produce the reverse of happiness."[12] Because pleasure and pain are consequences of actions it also follows that the rightness or wrongness of a course of action lies not in the intention that motivated it, but in the outcomes it produces; thus utilitarianism is in the class of *consequentialist* ethical theories, the spirit of which is often captured by the phrase, "the end justifies the means."

Of most importance to our discussion of the moral point of view, utilitarian ethical theory posits a decision procedure, the *hedonistic calculus,* according to which the moral agent can rationally determine right and wrong. The point of the calculus is to determine the relative proportion of "pleasurable" (good) and "painful" (bad) consequences associated with each potential response to the ethical dilemma under consideration; the moral agent then chooses that response which produces, on balance, "the greatest good for the greatest number" or at least less pain than any other alternative. Seven factors are to be considered in this quantitative approach of weighing the units of good and bad (sometimes referred to as "utils") produced by a decision:

1. intensity (how strong the pleasure or pain),
2. duration (how long into the future the pleasure or pain will last),
3. certainty (how likely it is that pleasure or pain will indeed result),
4. propinquity (how long it will be before the pleasure or pain is experienced),
5. purity (how likely it is that the pleasure will be followed by pain or the pain by pleasure),
6. fecundity (how likely it is that the pleasure will be followed by more pleasure or the pain by more pain),
7. extent (how many people will experience the pleasure or pain produced).[13]

Applying the general principle of "the greatest good" to the particular situation, the moral agent is to quantitatively assess all of the good and bad consequences associated with each course of action, whenever they may occur and whomever they may affect (this is, of course, an idealized procedure that few if any adherents actually live up to in practice, given the virtual impossibility of identifying all consequences much less accurately assigning relative weights). Note that performing this calculus is above all a function of reason in that it hinges on the logical assessment of consequences in rigorous mathematical terms; it also requires the decision-maker to remain impartial in that consequences to the self or to loved ones are to be weighted no differently than those to which one is neutral (in Bentham's words, "Each to count for one and none for more than one").

Deontological Ethical Theory

If Bentham's utilitarian ethical theory captures above all Cartesian rationalism's tendency toward the rigorously quantitative, it is the *deontological* theory of Immanuel Kant that embodies its emphasis on rationality as the defining human characteristic, on the consequent autonomy of the thinking individual, and on the ultimate authority of universal principles. The word deontology comes from the Greek "deon" for *duty*, and it is this concept that defines the essence of morality for Kant: rather than acting to maximize good consequences, we are to act always out of duty, which entails acting in accordance with universal principles.

Just as Bentham began with a theory of human nature (that we are ultimately driven by pleasure and pain) as the simple truth or first principle grounding utilitarianism, so too did Kant posit certain crucial facts about the human condition as the foundation of his philosophy. Not surprisingly given the Cartesian backdrop of late-eighteenth-century thought, Kant held that the human being was distinctly rational.

Also differing with Bentham and Mill on the question of goodness and thus grounding his ethical theory in an entirely different set of assumptions, Kant states clearly in the opening line of his *Groundwork of the Metaphysic of Morals* that the rational will alone is good: "It is impossible to conceive anything at all in the world, or even out of it, which can be taken as good without qualification, except a *good will*." Human beings have a rational capacity, a capacity for self-legislation and universal reasoning, that is often imperfectly expressed. Since we have the choice, then, between perfect and imperfect expression of the good will, since we can choose to be rational or not, we are autonomous moral agents: rational action has positive moral worth, and the imperfect expression of the will is moral error.

To the extent that we act purely from our rational will, rather than from such "inclinations" as desire, or love, or fear, we are acting from duty. Such action will, by the very definition of rationality, be in accordance with universal principles. Kant sums up the relationship between the rational will and universal principles in the *Categorical Imperative,* the core statement of his ethical theory: "I ought never to act except in such a way that I can also will that my maxim should become a universal law." We cannot, for example, will, without being self-defeating or inconsistent (and thus irrational) that everyone can make promises with the intention of breaking them "for by such a law there could properly be no promises at all, since it would be futile to profess a will for future action to others who would not believe my profession; . . . consequently my maxim, as soon as it was made a universal law, would be bound to annul itself." Making promises that I intend to break, or lying, is therefore a violation of the Categorical Imperative, is irrational, is a failure of duty, and is immoral (all roughly equivalent phrases for Kant). Because a universal law by definition permits no exceptions, lying is therefore wrong: categorically, universally, for any and all moral agents, period.

Note, however, that if I refrain from lying out of fear of getting caught or out of concern for another person, my behavior, while *in accordance with duty* is actually *out of "inclination"* not *out of duty* and thus has no moral worth. Moral deliberation, therefore, must eliminate inclination (emotions, desires, and so on) in favor of reasoned adherence to duty. Further, "an action done from duty has its moral worth *not in the purpose* to be attained by it, but in the maxim in accordance with which it is decided upon." The obviously nonconsequentialist result is that we do the right thing solely because it is the right thing to do, right being defined by the universal principles determined by the rationality of the Categorical Imperative. And, again, our autonomy lies in our individual capacity to reason our way to universal principles and in our choice to "legislate" or will adherence (or not) to these principles.

It is this very autonomy, in fact, that forms the basis of another important concept in deontological ethical theory, the concept of human *rights*. The nature of rational beings, according to Kant, "already marks them out as ends in themselves . . . whose existence has in itself an absolute value . . . — that is, as something which ought not to be used merely as a means—and consequently imposes to that extent a limit on all arbitrary treatment of them (and is an object of reverence)." The respect due our unconditional worth thus finds its expression in our having rights, which are essentially designed to prevent our being treated merely as means to others' ends, as if our value were not absolute but rather contingent on the ends or purposes of others.[14]

Rationalism in Contemporary Moral Philosophy

Utilitarianism has evolved since the time of Bentham and Mill, the emphasis on producing pleasurable consequences being modified to read in terms of "preference satisfaction," for example (such that the right course of action is that alternative which violates the fewest, or promotes the most, personal preferences). Many of its practitioners now argue that it is at the level of deciding among rules that one should apply the calculus rather than at the point of choosing among particular actions (thus "rule utilitarianism" advocates weighing the consequences associated with society having a rule that prohibits killing, while "act utilitarianism" enjoins performing the calculus every time one considers the act of killing another). Deontological ethical theory has also evolved from Kant's original writings. Some theorists, for example, have come to integrate a reasonable consideration of consequences into the Kantian decision-making process (although the intent to adhere to one's duty, rather than the consequences of one's actions, remains the determining factor). Paralleling utilitarianism, deontology has also been refined into "act" and "rule" versions. Elements such as the Categorical Imperative and the universal principles that are simultaneously discovered by and legislated by the rational will have come to be understood in less metaphysical terms as aspects of "moral realism": an understanding of morality that posits the existence of objective moral truths. As with the persistence of utilitarianism's ideal of a rigorous, quantitative decision procedure, the core of Kantian ethical theory remains: the autonomous moral agent is fundamentally a rational, impartial will acting out of duty to adhere to universal principles that render categorical judgments of right and wrong and operationalize respect for the inherent value of rational persons.

In many ways *ideal observer* theory is the twentieth-century incarnation of Cartesian rationalism and the MPV. Building on the evolution of ideas about legitimate moral judgment within utilitarian and deontological ethical theory, contemporary philosophy has begun to consider as a standard for the determination of moral goodness or rightness the judgments of a hypothetical person who explicitly embodies the traits required for legitimate ethical deliberation. The ideal observer has knowledge of all the relevant (nonmoral) facts; is impartial, disinterested, and dispassionate; has a commitment to consistent, logical reasoning; and is "normal" or like us in all other respects. This hypothetical construct helps to explain the sense that we all have from time to time of knowing that we "should" approve of X as a good thing even though in actuality we don't feel positively about it; we sense that the ideal observer *would* approve, and we recognize our own discrepant judgment as evidence of the failure of our process of ethical deliberation to live up to the standards of legitimacy (that is, we recognize that our judgment is being swayed by selfish or emotional considerations). In the eyes of many contemporary moral philosophers, ideal observer theory is the most morally attractive justification for ordinary moral judgments, insofar as we respond to the image with an affirming "Yes, that is what I meant by saying that X is good/right."

The premiere contemporary representative of the tradition of rationalism in moral philosophy is Harvard professor John Rawls. The relevance of his work, *A Theory of Justice,* for our discussion here lies in its depiction of a position from which we make legitimate decisions in our efforts to establish a just society. Rawls combines the concept of an appropriately situated observer with a contract approach to moral judgment in which "right" is equivalent to that which people would agree to on reflection and postulates the hypothetical "original position" from which ideal decisions about justice are to be rendered. Rawls writes, "The idea of the original position is to set up a fair procedure so that any principles agreed to will be just. . . . Somehow we must nullify the effects of specific contingencies which put men at odds and tempt them to exploit social and natural circumstances to their own advantage." His method is to place the deliberators behind a "veil of ignorance" and thus ensure that they will approximate the ideal observer: although they are all equally rational, from behind the veil they "do not know how the various alternatives will affect their own particular case and they are obliged to evaluate principles solely on the basis of general considerations." When judging from behind the veil, the hypothetical people know nothing of the particulars of their situations in the real world: they do not know their "place in society, [their] class position or social status," their "fortune in the distribution of natural assets and abilities . . . intelligence, strength, and the like," their "conception of the good, the particulars of [their] rational plan of life, or even the special features of [their] psychology such as [their] aversion to risk or liability to optimism or pessimism." The perspective of the original position, then, is Rawls's version of the MPV.[15] The rationalist ideal remains: the closer we approximate this appropriately situated ideal observer, eliminating from our deliberation process all the particular details of the situation that might serve to bias our thinking while remaining committed to logical reasoning from a comprehensive knowledge base, the sounder and more legitimate our moral reasoning will be.

Perhaps the most significant development in contemporary rationalist tradition philosophy, however, has been the extension of this perspective beyond the human realm and into the environmental realm. In the remainder of this chapter we will explore the environmental ethics theorizing that has grown out of the mainstream rationalist tradition. The *extensionist project,* as it is called, is modeled quite closely on traditional philosophy; therefore, it embodies many of the same assumptions about ethical deliberation that we undertook to make explicit above. Although they do not always refer to the moral point of view by name, extensionist authors implicitly assume the adoption of an impersonal and impartial perspective in ethical decision-making and prescribe the rational application of abstract universal principles; their theorizing is in many ways hierarchical and dualistic, and it is oriented toward the resolution of competing claims and interests. Indeed, the primary difference between their thought and that of such rationalist philosophers as Mill or Kant is the broader scope they envision for the application of these principles and concepts, a scope that varies from author to author but that, by definition, always includes at least some nonhumans.

As we successively turn our attention to the work of each of the major environmental ethicists, let us take care to keep in mind the core rationalist concepts as we have begun to identify them; we will revisit the defining characteristics of this tradition at the conclusion of this chapter, examining in detail their influence on the evolution of moral philosophy within this framework. Here and in the exploration of the two alternative frameworks in Parts II and III, awareness of the parallels and shared assumptions between the traditional philosophical canon and the environmental ethics theorizing that has emerged from it will serve two important purposes: it will enhance our appreciation of the evolution of philosophical thought within each tradition, and it will strengthen our ability to engage in critical comparison of the perspectives on environmental ethics offered by our three frameworks. We begin with an overview of the challenge extensionism poses to traditional philosophy, and we then explore each of the three major approaches within extensionism: animal welfare, biocentrism, and ecocentrism.

The Extensionist Impetus

Although we did not explicitly discuss this aspect of their thought earlier in the chapter, a careful look at the theorizing of Descartes, Bentham, and Kant reveals that philosophers have long considered the question of the moral status of nonhumans. Animals, because they lacked a rational "soul," were mere "machines" to Descartes; and in Kant's theorizing they were given the status of "things," which we have an indirect duty to treat well only out of respect for other people or out of fear of developing, through habitual mistreatment of them, a harsh or mean-spirited character. Bentham, however, posited the future inclusion of animal pain and pleasure in the hedonistic calculus in a passage from *Principles of Morals and Legislation* that anticipates the extensionist project:

> The day *may* come when the rest of the animal creation may acquire those rights which never could have been withholden from them but by the hand of tyranny. The French have already discovered that the blackness of the skin is no reason

why a human being should be abandoned without redress to the caprice of a tormentor. It may one day come to be recognized that the number of the legs, the villosity of the skin, or the termination of the *os sacrum,* are reasons equally insufficient for abandoning a sensitive being to the same fate. What else is it that should trace the insuperable line? Is it the faculty of reason, or perhaps the faculty of discourse? But a full-grown horse or dog is beyond comparison a more rational, as well as a more conversable animal, than an infant of a day, or a week, or even a month old. But suppose they were otherwise, what would it avail? The question is not, Can they *reason*? nor Can they *talk*? but, Can they *suffer*?[16]

The difference underlying the thinking about animals in these examples has come to be expressed in terms of *moral standing*: for Descartes and Kant the welfare of animals did not have to be taken into account in the process of ethical deliberation and they thus lacked moral standing, whereas Bentham anticipated the future understanding of our treatment of animals as subject to ethical constraints or our understanding of them as *moral subjects* with moral standing. Having moral standing or being a moral subject is equivalent to being a member of the *moral community*, the realm of application of ethical considerations. Limiting the moral community to human beings, asserting that we alone are morally considerable, is *anthropocentric* or human-centered, while acknowledging the moral standing of at least some nonhumans is *nonanthropocentric*. Under either of these paradigms, normal, adult human beings are both moral subjects and *moral agents*: rational deliberators who define problems, make decisions, and initiate action accordingly and who can be held accountable for the integrity and consequences of their choices. The project of extensionism involves expanding the moral community to include at least some nonhumans or, in other words, acknowledging their moral standing; extensionism challenges as arbitrary, biased, and irrational the long and widely held anthropocentric assumption that humans alone are morally considerable subjects. And as the Bentham quote above suggests, this extension of moral considerability parallels historical progressions in our ethical thinking in the human realm and similarly holds the potential for drastically transforming our treatment of nonhumans.

A passage from Aldo Leopold's *A Sand County Almanac* has been widely, though not uncontroversially, read as the initiating call to extensionism in moral philosophy, and Leopold has thus come to be known as the "father of environmental ethics."[17] Leopold was disturbed by the overly instrumental view of the environment demonstrated by resource managers and conservationists alike, a view characterized with the Abrahamic image: "Abraham knew exactly what the land was for: it was to drip milk and honey into Abraham's mouth." He critiques the mindset in which appeals for the protection of nature can only be expressed in terms of self-interest because "it defines no right or wrong, assigns no obligation, calls for no sacrifices, implies no change in the current philosophy of values," and he writes: "The 'key-log' which must be moved . . . is simply this: quit thinking about decent land-use as solely an economic problem. Examine each question in terms of what is ethically and esthetically right, as well as what is economically expedient." Thus what distinguishes Leopold as the initiator of a new, specifically environment-oriented type of ethical theorizing is his understanding of environmental problems as inappropriately subject to an economic rather than an ethical value system and his call for a broadening of the realm of ethical concern to include the natural world.

The passage that articulates the impetus toward extensionism is an analogy for expanding the moral community embodied in the story of Odysseus:

> When god-like Odysseus returned from the wars in Troy, he hanged all on one rope a dozen slave-girls of his household whom he suspected of misbehavior during his absence. This hanging involved no question of propriety. The girls were property. The disposal of property was then, as now, a matter of expediency, not of right and wrong. . . . The ethical structure of that day . . . had not yet been extended to human chattels. During the three thousand years which have since elapsed, ethical criteria have been extended to many fields of conduct, with corresponding shrinkages in those judged by expediency only. . . . There is as yet no ethic dealing with man's relation to land and to the animals and plants which grow upon it. Land, like Odysseus' slave-girls, is still property. The land-relation is strictly economic, entailing privileges but not obligations.

Thus, while ethical structures have been extended to the point that girls can no longer be considered as disposable property (meaning that ethical considerations, not questions of expediency, have become the determining factors regarding their treatment), "land," Leopold's collective term for "soils, waters, plants, and animals," still has the status of Odysseus's slave-girls: it is not included in the moral community and is therefore not directly taken into account in ethical deliberation. Plants, animals, and the environment as a whole are treated as expediency or efficiency or desire dictates. And just as the slave-girls died as a consequence of being excluded from the moral community, so too does the natural world suffer from our failure to grant it the status of legitimate moral subject; consider as evidence, Leopold asks, "the soil, which we are sending helter-skelter downriver . . . the waters, which we assume have no function except to turn turbines, float barges, and carry off sewage . . . the plants, of which we exterminate whole communities without batting an eye . . . [and] the animals, of which we have already extirpated many of the largest and most beautiful species."

In stark contrast to this conceptualization of the natural world as lying beyond the realm of ethical concern, however, the core of Leopold's thinking and of extensionist ethical theory is that "the despoliation of land is not only inexpedient but wrong." Thus, our treatment of nonhumans and/or the natural world at large ought to be subject to ethical constraint; moral considerability ought to be extended beyond our own species just as it has historically been extended within it. Although the ethical theories of those extensionist philosophers who have developed and refined this original impetus sometimes take drastically different forms, arguing for the extension of moral considerability to different categories of nonhumans and adhering to different philosophical foundations, they all share Leopold's commitment to extending the realm of ethical concern beyond the presumably impermeable boundary of the human species. It is this approach that gives the "extensionist project" its name, its fundamental impetus, and its coherence.[18]

As we will see in later chapters, there are other approaches to environmental philosophy that do not rest on the premise that the next logical (and necessary) steps in the evolution of ethical theory involve continued expansion of the moral community. The extensionist premise has, however, been the foundation of mainstream environmental ethics theorizing and it makes historical sense that this be so, for the most obvious starting point of any new theorizing is the point at which the older theory leaves off. We turn now to an examination of the work of several of the primary extensionist authors, categorized according to the extent to which they expand the moral community.

Animal Welfare: Singer and Regan

Singer and Animal Liberation

Australian-born and Oxford-educated, Peter Singer has written in the field of political philosophy and on the ethics of poverty and hunger; the work of concern to us here is his 1975 text *Animal Liberation: A New Ethics for Our Treatment of Animals*. Its simple yet profound thesis is stated in the opening paragraph of the Preface:

> This book is about the tyranny of human over nonhuman animals. This tyranny has caused and today is still causing an amount of pain and suffering that can only be compared with that which resulted from the centuries of tyranny by

white humans over black humans. The struggle against this tyranny is a struggle as important as any of the moral and social issues that have been fought over
in recent years.[19]

These lines clearly reveal both Singer's extensionism and his utilitarianism.
His is a call for the moral community to expand once again to include nonhuman animals, as it previously did to include racial minorities, and it is a
call grounded in condemnation of the unjustified infliction of pain on those
beings otherwise beyond the reach of ethical concern.

Singer thus picks up utilitarian ethical theory where it left off and extends
its range into the nonhuman realm. The core of Singer's work concerns the
inconsistency of the traditional restriction of this theory to human pleasure
and pain: "Pain is pain, and the importance of preventing unnecessary pain
does not diminish because the being that suffers is not a member of our
species."[20] Singer points out that if any nonhumans experience pleasure or
pain, then utilitarianism restricted to human beings is internally incoherent,
a loophole as it were that only an irrational, inconsistent reading of the theory could admit. The hedonistic calculus we are to perform when making
ethical judgments must take any and all nonhuman pleasure and pain into
account as well as human pleasure and pain.

Sentience, then, or the ability to experience pleasure or pain, not membership in the species *Homo sapiens*, demarcates the boundary of the moral community from the perspective of extended utilitarianism. The question then
becomes one of determining which if any nonhumans do indeed experience
pleasure or pain or, more generally, whether the boundary around the human
species coincides with the boundary around potential experiencers of pleasure or pain. Given similar behavioral signs, similar nervous systems and
physiological responses to stimuli, and a common evolutionary experience of
the adaptiveness of sensitivity to pain, we can be fairly confident attributing
sentience and thus membership in the moral community, to animals with a
sufficiently developed nervous system; and we can exclude plants, fungi, mi-
croorganisms, and, of course, nonliving objects. Thus, mammals and birds
are definitely "in," fish and reptiles are probably "in," and insects and most
mollusks (oysters, clams, and so on) are probably "out."[21]

Sentient animals have interests, at the very least the interest in avoiding
pain. And it is consideration of *interests* that lies at the heart of utilitarian ethical theory: "The moral basis of equality among humans is not equality in fact
[which Singer suggests would be an absurd claim given our obvious inequality with respect to such characteristics as intelligence and physical strength],
but *the principle of equal consideration of interests*, and it is this principle that,

in consistency, must be extended to any nonhumans who have interests."[22] Thus, "the interests of every being that has interests are to be taken into account and treated equally with the like interests of any other being."[23]

It is important to note here that, just as in the human realm of ethics, equal consideration of interests in an expanded moral community does not imply identical treatment or an inability to choose between conflicting interests. If we must choose between saving the life of a dog or that of a human being, it may be that the person's interests—given his awareness of and fear of death, the sorrow his family will experience upon his death, and the potential for a long and happy life—outweigh those of the dog; thus, the principle of equal consideration of interests tells us to save the person rather than the dog, not because we discount the dog as a nonhuman but rather because the balance of interests tilts toward the human in this case. In another case—for instance one in which the human being would not suffer as intensely (not fearing death perhaps or being mentally impaired) and has no one to grieve for her— the weighing of interests could conceivably favor the dog.[24] Making such comparisons is, of course, difficult, but although extended utilitarianism is further complicated in this respect, we encounter the same problem in interhuman ethics.

The point of the foregoing example is that in extended utilitarianism, we cannot justify choices made in accordance with the morally arbitrary line of species membership; rather it is the morally relevant line of interests that must guide our deliberation. Choosing to save the person over the dog *because he is human and the dog is not* or justifying the use of animals but not people for food, clothing, or sport *because they are not human* constitutes *speciesism:* "a prejudice or attitude of bias toward the interests of members of one's own species and against those of members of other species."[25] Bentham's "The day may come" passage draws the parallel to racism that the term is intended to evoke; just as racial membership has long, but wrongly, served to justify differential treatment of minorities, so too does species membership inappropriately justify distinctions between ourselves and other animals, which lead us to treat them in ways we would not consider treating each other. If the relevant consideration is suffering, then whether that suffering is experienced by a dog or by a human is just as meaningless a distinction as whether it is experienced by an African American or by a Caucasian American. A similar parallel can and has been drawn with sexism, the unjustified differential treatment of men and women based solely on the distinction of sex. Of course, racism, sexism, and speciesism are not unethical because they posit differences (which it would be foolish to argue do not exist across races, sexes, or species) but rather because they take as legitimating differential treatment differences that are morally irrelevant.

What, then, are the implications of extending the moral community to include all sentient animals, human or nonhuman? Singer focuses most of his discussion of the practical consequences of "animal liberation" on two issue areas: animal experimentation (for academic, medical, and commercial purposes) and factory farming (the industrial raising and processing of animals for human consumption). To judge the morality of any form of animal experimentation on Singer's terms, we must perform a hedonistic calculus that includes all interests, human and nonhuman alike. Thus if the issue is one of testing a potentially cancer-curing agent on chimpanzees, the human benefits of the medical advance (considering the numbers of people afflicted with cancer and the pain they suffer) could conceivably outweigh the suffering experienced by the test animals, particularly if the pain involved and the number of subjects are minimized; in such a case, Singer's theory would approve the animal experimentation as the correct choice. In contrast, painful and often repetitive tests in the cosmetics industry that are designed to assess the safety of the sixty-first shade of pink lipstick (for example) may very well involve such great suffering for such inconsequential human gains as to be unjustifiable. We therefore have to take each instance of animal experimentation on a case-by-case basis and carefully weigh all the interests that are at stake, ultimately judging as right that choice which, on balance, produces less net pain (fewer net negative consequences) than any other option. With the bias toward our own interests removed and given the suffering it necessarily entails, animal experimentation is prima facie unethical (wrong, unless an overriding consideration can be brought to bear) and can be justified only when there is no other, less painful, option for achieving crucial human gains; and where it is justified, such testing must minimize the suffering involved.

The same dynamic holds for factory farming, although this is a more clearcut case. Eating meat is a luxury, necessary neither for our survival nor our health. Insofar as the consumption of meat and other animal products is a pleasurable experience, this enjoyment must be taken into consideration; also to be weighed in, however, is the intense suffering experienced by animals in the process. With factory farming as with cosmetics testing, the human gains are trivial and do not come even close to offsetting the misery experienced by the animals:

> Since . . . none of these practices [penning chickens in overcrowded cages, immobilizing calves raised for veal, etc.] . . . cater for anything more than our plea-

sures of taste, our practice of rearing and killing other animals in order to eat them is a clear instance of the sacrifice of the most important interests of other beings in order to satisfy trivial interests of our own . . . we must stop this practice, and each of us has a moral obligation to cease supporting the practice.[26]

Granting that many conflicts between human and nonhuman interests are probably more complex and thus less amenable to clear-cut resolution than is the question of meat-eating, Singer nevertheless holds that his extended utilitarianism has significant and wide-ranging implications for our treatment of nonhumans. "What we must do," he writes, "is bring nonhuman animals within the sphere of moral concern and cease to treat their lives as expendable for whatever trivial purposes we may have."[27] Singer's explication of the logic (or rather illogic) of speciesism—"the claim that to discriminate against beings solely on account of their species is a form of prejudice, immoral and indefensible in the same way that discrimination on the basis of race is immoral and indefensible"[28]—serves as the basis for many subsequent extensionist critiques of anthropocentrism.

Regan and Animal Rights

Tom Regan is not infrequently referred to as the "intellectual leader of the animal rights movement," and his text, *The Case for Animal Rights,* has come to serve as its philosophical underpinning. Regan, like Singer, offers a rigorously defended argument against viewing nonhumans as mere resources whose treatment is not directly subject to ethical constraints. Unlike animal liberation, however, the goals of animal rights (at the level of both philosophical theory and popular movement) are categorical: "the *total abolition* of the use of animals in science, the *total dissolution* of commercial animal agriculture, and the *total elimination* of commercial and sport hunting and trapping."[29] Regan's extensionism is not utilitarian in orientation (requiring the weighing of interests) but rather *deontological* (articulating universal principles of right and wrong which we are duty-bound to follow); it is the tradition of Kant that Regan extends beyond the traditionally human realm.

The core of Regan's theory, then, closely parallels the familiar deontological pattern:

$$\text{inherent value} \rightarrow \text{respect} \rightarrow \text{rights.}$$

Our inherent value, which we all possess equally, is to be acknowledged with respect, and we are thus not to be treated "merely as means" to others' ends;

such treatment is a violation of our rights as individuals. Regan disagrees, however, with Kant's linking the inherent value of human beings to our rationality, on the grounds that the theory itself is internally inconsistent and arbitrary. Since not all humans have a rational capacity (infants, the mentally defective, the insane, and the comatose, for example), the theory does not actually offer a basis for acknowledging the inherent value of human beings as such but rather technically, and arbitrarily, selects only those human beings who are moral agents. The wrongness of harming an infant (a human who does not happen to be a moral agent) does not lie in the indirect affront to her parents (who are moral agents) but rather in the direct injury to her; she has the same claim to respect as does an adult, fully rational, human agent, and thus we have the same duty to avoid harming her. Rationality is thus not a legitimate criterion for the possession of inherent value, even within the realm of human ethics; and since the criterion of membership in the biological species *Homo sapiens*—the only criterion that would select all and only human beings—constitutes "blatant speciesism" for Regan as for Singer,[30] the door is thus opened for a modified deontological discussion of inherent value, respect, and rights that is not limited to the human realm.

If it is not by virtue of our rationality or our species membership that we legitimately acknowledge one another as inherently valuable, then what is the basis of such assertions? The most basic and morally relevant characteristic shared by the noncontroversial, paradigm cases of inherent value in human beings, Regan concludes, is the "subject-of-a-life criterion":

> We are . . . each of us a conscious creature having an individual welfare that has importance to us whatever our usefulness to others. We want and prefer things; believe and feel things; recall and expect things. And all these dimensions of our life, including our pleasure and pain, our enjoyment and suffering, our satisfaction and frustration, our continued existence or our untimely death—all make a difference to the quality of our life as lived, as experienced by us as individuals.[31]

Although we may vary in the degree to which we may be said to hold these characteristics, the status of "subject-of-a-life" does not in itself admit of degree: "one either *is* a subject of a life, in the sense explained, or one *is not*. All those who are, are so equally." And, of course, the same is true of the corollary attribution of inherent value: "one either has it, or one does not. There are no in-betweens. Moreover, all those who have it, have it equally. It does not come in degrees."[32]

Normal mammals at least one year of age are, according to Regan, uncontroversially selected by this criterion. The dog, Fido, who can serve as Regan's representative "subject of a life that is experientially better or worse," not only shares our physiology and our mammalian evolutionary history (and thus our sentience and consciousness); he also behaves in a nonrandom and predictable fashion in response to an environment of which he is intimately aware; he selects a favored food when presented with a range of options (indicating preferences, desires, learning, memory, and so on); he believes that a friend is at the door upon hearing familiar footsteps, anticipates her arrival, and experiences disappointment when she retreats without coming inside; and he initiates actions designed to bring about desired future consequences (barking at the door in order to go to the backyard, for example, indicating autonomy, intentionality, and self-consciousness). In short, Fido is the subject-of-a-life that fares well or ill for him; he, as well as all relevantly similar animals, has inherent value and must be acknowledged as a member of the moral community.[33]

For Kant, respect for our absolute value as rational beings means refraining from treating each other "merely as means," as if our value were contingent on our usefulness to others; our specific rights, then, protect us from the various ways in which we could be devalued in this manner. Regan's theory similarly moves from the acknowledgment of inherent value to the injunction of respectful treatment. His *respect principle* states, "We are to treat those individuals who have inherent value in ways that respect their inherent value;" and it explicitly includes the duty to refrain from harming an individual with inherent value "merely to produce the best consequences for all involved" (which explicitly treats that individual as if his value were a function of his utility to others) and the duty to come to the assistance of those individuals with inherent value who have been treated with disrespect by others. And his *harm principle* states, "we have a *prima facie* direct duty not to harm those individuals who have an experiential welfare"; since subjects-of-a-life "are individuals who have an experiential welfare—whose experiential life fares well or ill, depending on what happens to, or is done to or for, them, . . . we fail to treat such individuals in ways that respect their value if we treat them in ways that detract from their welfare—that is, in ways that harm them." (Although similar to the first of the two duties of the respect principle, the harm principle is broader; the first duty of the respect principle, in effect, specifies that producing the best consequences for all concerned is not sufficient cause for overriding the harm principle.) Regan, like Kant, makes it clear that such respectful treatment is something individuals with inherent value are *owed;* in other words, human and nonhuman subjects-of-a-life all have an equal right to be treated with respect for our inherent value, and any specific rights we or they may have protect us from particular ways in which this fundamental right might be violated.[34]

The right to respectful treatment does not imply, any more than did equal consideration of interests, an inability to make difficult choices or to sacrifice one specific right for another. Recall that Singer's extended utilitarianism, although requiring that we invoke the principle of equal consideration of interests, nevertheless justified sacrificing a dog to save a person in some cases while sacrificing the person to save the dog in others. Is there a corresponding way to decide whether the dog or the person lives when both are acknowledged as having equal inherent value and the equal right not to be used as a means to the others' ends? Regan addresses just this dilemma in his discussion of a hypothetical lifeboat that can only hold four out of five survivors: four normal adult humans and one adult dog. Although the theory of animal rights requires that we not sacrifice the dog *because he is a dog and not a person* and that we not violate his right to equal respect, it also enjoins us to make such choices in ways that acknowledge unequal harm; and since the harm involved in the dog's death ("a function of the opportunities . . . it forecloses") is understood to be less than the harm involved in the death of a human being, the dog's prima facie right not to be harmed is justifiably overridden and he is sacrificed in order to save the people.[35]

Unlike Singer's utilitarianism, Regan's animal rights theory is categorical or absolute in its condemnation of such practices as factory farming, sport hunting, and animal research because each amounts to using animals as means to our ends. In arguing against factory farming, for example, Regan acknowledges the infliction of pain, the frustration of physical movement, and the other conditions that render the animals' lives miserable not as the fundamental wrong but as "symptoms of the deeper systematic wrong that allows these animals to be viewed and treated as lacking independent value, as resources for us." As is also the case with using animals for experimental or research purposes, "giving [them] more space, more natural environments, more companions . . . [or] more anaesthesia or bigger, cleaner cages . . . does not right the fundamental wrong"; and neither can either practice be justified by reference to the "real promise of human benefit." Since the fundamental wrong underlying all of these practices involves treating animals "routinely, systematically, as if their value is reducible to their usefulness to others, . . . they are treated with a lack of respect, and thus are their rights . . . violated." Just as we categorically condemn eating, experimenting on, and hunting other human beings on these grounds, regardless of the pleasure or knowledge gains that such activities might produce, so too is the only justifiable response here the absolute refusal to condone the fundamentally disrespectful treatment of animals on which these three practices are based. Animals are not resources, and to sanction or participate in the systematic violation of their rights inherent in factory farming, sport hunting, and experimentation is thus a failure in our duty to act only in ways that respect their inherent value and to defend their rights to such respectful treatment.[36]

Regan's development of the extensionist impetus thus differs significantly from Singer's approach. His subject-of-a-life criterion for moral considerability selects a smaller set of nonhumans for (certain) inclusion in the moral community but secures for them the protection from merely instrumental valuation, which is corollary to the possession of rights. The subject-of-a-life criterion does not select all humans (think here of the fetus and the comatose as representative problematic cases) or all nonhumans, but it is offered as a conservative demarcation that is well within the realm of certain moral standing. "The subject-of-a-life criterion is set forth as a *sufficient, not as a necessary, condition* of making the attribution of inherent value intelligible and nonarbitrary,"[37] writes Regan, thus not ruling out the possibility of birds, fish, flowers, or even rivers having inherent value; and, in fact, he acknowledges that the establishment of a truly comprehensive environmental ethic hinges on this very attribution of inherent value to natural objects and nonmammalian nonhuman organisms. What is most problematic in any such discussion of the potential limits of moral considerability is *human chauvinism*: "a failure or refusal to recognize that those characteristics one finds most important or admirable in one's self, or in members of one's group, are also possessed by [other] individuals."[38] The term touches on the very heart of anthropocentrism in that it is our presumption that inherent value is a category that can apply only to ourselves as human beings that justifies our treatment of nonhuman animals as mere means to our ends.

Biocentrism: Taylor and Goodpaster

Taylor and Respect for Nature

Taylor's extensionism parallels in many ways the Kantian concept of respect for persons but is more inclusive in its scope than either Regan's extension of inherent value to animals meeting the subject-of-a-life criterion or Singer's extension of equal consideration of interests to animals meeting the criterion of sentience. Taylor clarifies the deontological underpinnings of his theory early in his text, *Respect for Nature: A Theory of Environmental Ethics:*

> The natural world is not there simply as an object to be exploited by us, nor are its living creatures to be regarded as nothing more than resources for our use and consumption. On the contrary, wild communities of life are understood to be deserving of our moral concern and consideration because they have a kind of value that belongs to them inherently. Just as we think it inappropriate to ask, What is a human being good for? because such a question seems to assume that the value or worth of a person is merely a matter of being useful as a means to some end, so the question, What is a wilderness good for? is likewise considered inappropriate. . . . The living things of the natural world have a worth that they possess simply in virtue of their being members of the Earth's Community of Life. . . . Just as humans should be treated with respect, so should they.

Taylor thus posits duties to nonhumans on the basis of their inherent worth as ends in themselves. (To head off a confusion, let us note at this point that where Kant and Regan refer to "inherent value," Taylor prefers the term "inherent worth" since he views it as less suggestive of the need for a valuer—the important point is just to keep in mind how each author uses the terms without becoming confused by the inconsistency from one author to the next.)

As we saw in Regan's theorizing, such extension does not link inherent worth to the possession of rationality but rather to characteristics that are deemed to have greater moral relevance, and in Taylor's theory the link to inherent worth is even less restrictive than the subject-of-a-life criterion: "Moral subjects must be entities that can be harmed or benefitted." The theory of "respect for nature" thus claims membership in the moral community for *all animals and all plants*; because they are alive their well-being can either be promoted or hindered, and "it is possible for us imaginatively to look at the world from their standpoint, to make judgments about what would be a good thing or a bad thing to happen to them, and to treat them in such a way as to help or hinder them in their struggle to survive." Conversely, inanimate objects cannot be so affected and thus do not have the status of moral subjects. Dogs and butterflies, daisies and oaks are "in"; rocks and rivers, canyons and sand dunes are "out." Thus, if it is unethical to dam a river, the wrongness of the act resides not in the mistreatment of the river (which can be neither harmed nor benefited) but rather, for instance, in the harm inflicted on the fish who live in the river.

In another important distinction between his theorizing and that of either Singer or Regan, Taylor explicitly eliminates from his theory consideration of those nonhumans who are not fundamentally allied with the "natural ecosystems" or the "populations of animals and plants that make up the biotic communities of those ecosystems." The concerns on which Singer and Regan focus much of their attention here fall under what Taylor calls "the ethics of the bioculture": a third ethical system (in addition to respect for persons and respect for nature) that is "concerned with the human treatment of animals and plants in artificially created environments that are completely under human control." Issues that do fall within the realm of respect for nature involve nonhumans as they are found in natural rather than artificial systems: hunting, fishing, and trapping; habitat destruction; oil spills; predator control; waste disposal; ecosystem and species preservation; and ecological restoration, among others.

Having determined what is unique about Taylor's approach to extensionism and how he demarcates the moral community, we are now in a position to briefly explore the three components of his ethical theory.

The attitude of respect for nature
This first component involves recognizing that beings with a good-of-their-own have inherent worth. An entity has a good-of-its-own if its well-being can be promoted or hindered solely in reference to itself. Thus a car does not have a good-of-its-own because it cannot be understood to be harmed independently of its having an owner whose purposes for the car are frustrated by our treatment of it; conversely, a tree does have a good-of-its-own since the harm that comes from insufficient water or nutrients involves the tree itself, whether or not any person has an interest in its well-being. All animals and all plants can be harmed or hindered in and of themselves, thus they all have a good-of-their-own; adopting the attitude of respect for nature involves recognizing that they thus have inherent worth, and acknowledging their inherent worth, we deem them worthy of respect or moral consideration. Once we have come to regard all wild animals and plants as having inherent worth (once we have adopted the attitude of respect for nature) we grant them equal status as moral subjects, we refrain from treating them merely as means to our ends (which would have the effect of reducing their worth from absolute to contingent), and we adopt as a moral principle the consideration and promotion of (or at least the noninterference with) their good for their own sake.

The biocentric outlook on nature
This second component of Taylor's ethical theory is the belief system or way of looking at nature and our place in it that undergirds and justifies the attitude of respect, and it consists of four beliefs:

1. "Humans are members of the Earth's Community of Life in the same sense and on the same terms" as other living things, with similar survival requirements, a joint need to be free of constraints if we are to flourish, and a shared origin in evolutionary processes.
2. The natural world is an interdependent system in which the survival and flourishing of each organism is integrally linked with that of others, the human species included.
3. Every organism is a "teleological center of life," an individual existence that has its own unique way of preserving itself and responding to its environment, and we thus can view it as an irreplaceable individual and look at the world from its point of view.
4. Humans are not inherently superior to other forms of life, and species-impartiality is demanded by the recognition that all organisms, human and nonhuman, have a good of their own and thus the same inherent worth.

The ethical system or rules of conduct

The third component of Taylor's ethical theory is a set of four primary rules that will be followed by any moral agent who has adopted the attitude of respect for nature and the biocentric outlook. These rules specify prima facie duties; they tell us what we should or should not do with respect to the natural world in the absence of competing claims, and of course often there will be overriding considerations that justify our breaking the rules.

1. *The rule of nonmaleficence.* We have a duty to avoid harming entities with a good of their own, including not killing them and not engaging in actions that are seriously detrimental to them.

2. *The rule of noninterference.* We have a duty to avoid restricting the freedom of organisms (either by the deliberate imposition of constraints—as in trapping—or by the alteration of the environment such that constraints are then imposed indirectly—as in filling in a lake that has formerly been used as a water source, because the absence of water is a constraint). And we have a duty to "acknowledge the sufficiency of the natural world to sustain its own proper order" and thus to adopt a general "hands-off" policy (not removing plants or animals from the wild—even if they subsequently would enjoy longer and healthier lives—and not managing or interfering with the natural functions of ecosystems).

3. *The rule of fidelity.* We have a duty to "uphold animals' expectations," to avoid deceiving, misleading, or otherwise breaking faith with them, espe-

cially with intent to harm them (including encouraging their trust or similarly camouflaging our intent to harm or kill them as in sport hunting).

4. *The rule of restitutive justice.* We have a "duty to restore the balance of justice between a moral agent and a moral subject when the subject has been wronged by the agent . . . [as when] an agent has broken a valid moral rule . . . [by making] amends [through] some form of compensation or reparation" (including compensating the population of an organism we killed and returning an organism we have harmed to its original condition or compensating for its reduced ability to flourish, perhaps by augmenting its habitat to ease access to food). Because "our well-being is constantly being furthered at the expense of the good of the Earth's nonhuman inhabitants," we all owe the natural world compensation and thus "should share in the cost of preserving and restoring some areas of wild nature for the sake of the plant and animal communities that live there."

Taylor also offers a hierarchy or ranking to which we can appeal in instances of internal conflict among the duties specified by these rules. For example, where possible, we must make restitution in ways that do not require harm, interference, or breaking faith, but when this cannot be achieved restitutive justice is to take precedence over noninterference and fidelity. Fidelity can similarly override noninterference, but nonmaleficence can never justifiably be violated since "our most fundamental duty toward nature is to do no harm to wild living things as far as this lies within our power." Thus, we can make restitution to an endangered species through limited captive-breeding programs, and we can compensate for an oil spill by cleaning the fur and feathers of affected animals (in both, potentially violating fidelity and noninterference); and we can restrict the freedom of movement within a woodland in which animals have come to feel safe by setting up a fence around a new hazard such as a garbage dump (adhering to the rule of fidelity by violating the rule of noninterference). We cannot, however, "commit a further wrong in an attempt to make up for a past wrong" by protecting the remnants of an endangered species through killing its predators, even if this is the only way compensation can possibly be made.

Finally, if humans and nonhumans alike are recognized as moral subjects, then such issues as clearing land for the construction of a hospital or art museum and plowing under a field of wildflowers in order to plant food crops present difficult moral dilemmas: "the conflict between the good of other species and the realization of human values . . . then appears to us as a situation of *competing moral claims.*" Taylor offers five principles for establishing priorities among competing claims between human ethics ("respect for persons") and environmental ethics ("respect for nature"):

1. *The principle of self-defense.* Killing in self-defense is permissible but, since we may inflict only the minimum necessary harm, only in the absence of all other alternatives; and we must take precautions to avoid being confronted with such a choice in the first place.

2. *The principle of proportionality.* Basic nonhuman interests ("whose fulfillment is needed by an organism if it is to remain alive") are to take precedence over nonbasic exploitative human interests (interests not related to survival, which reduce the value of the nonhuman in question to the merely instrumental). Hunting and fishing for sport and killing exotic animals in order to sell their horns or fur for trinkets or clothes are thus not permitted, "for all such practices treat wild creatures as mere instruments to human ends, thus denying their inherent worth."

3. *The principle of minimum wrong.* Nonbasic nonexploitative human interests (those that are not fundamentally incompatible with respect for nature and that are so important as to be not easily relinquished) may take precedence over basic nonhuman interests (Taylor assumes that they can only have basic interests), if they are met in such a way as to absolutely minimize the harm to the nonhumans. Constructing hospitals, art museums, airports, highways, or public parks at the expense of natural habitat and damming rivers for hydroelectric power are permitted even though organisms are harmed since, in such cases, "wild animals and plants are not being used or consumed as mere means to human ends."

4. *The principle of distributive justice.* When basic nonhuman interests conflict with basic human interests there must be "a just distribution of interest-fulfillment among all parties . . . [and] . . . each party must be allotted an equal share [of any] . . . natural source of good that can be used for the benefit of [either]." Subsistence hunting is allowed where human beings would otherwise starve, but generally vegetarianism is required since it reduces our land use and thus frees up land for nonhumans; similarly, we are obliged to preserve at least certain areas of land in an attempt to balance our resource use with that of nonhumans. We must try to "make it possible for wild animals and plants to carry on their natural existence side by side with human cultures," even while we realize that a perfectly just (equal) balance is virtually unobtainable.

5. *The principle of restitutive justice.* When either the Principle of Minimum Wrong or the Principle of Distributive Justice has been invoked we have justified a gain to ourselves at the expense (albeit minimized) of the natural world and in so doing we have incurred a duty of compensation. Wilderness preservation or "[setting] aside habitat areas so that wild communities of animals and plants can realize their good is the most appropriate way to restore the balance of justice with them."

Taylor has thus given us a "unified and comprehensive vision" of a world in which the moral community is extended to include all living beings in nature, and his most significant contribution in this project is perhaps his development of specific guidelines for the resolution of the increased moral conflict between humans and nonhumans, which such an egalitarian world necessarily entails. In his theorizing as in that of Singer and Regan, an "inner change in our moral beliefs and commitments" with respect to the natural world is the necessary precursor to the envisioned economic, legal, and political changes that will institutionalize an expanded moral community and the behavior changes that are subsequently prescribed. This inner change from anthropocentrism to nonanthropocentrism is not thought to be easily accomplished, but it is thought to be possible, and Taylor concludes *Respect for Nature* with an eloquent statement of this shared extensionist assumption:

> There should be no illusions about how hard it will be for many people to change their values, their beliefs, their whole way of living. . . . Psychologically, this may require a profound moral reorientation. Most of us in the contemporary world have been brought up in a thoroughly anthropocentric culture in which the inherent superiority of humans over other species has been taken for granted. Great efforts will be needed to emancipate ourselves from this established way of looking at nonhuman animals and plants. But it is not beyond the realm of practical possibility. Nothing prevents us from exercising our powers of autonomy and rationality in bringing the world as it is gradually closer to the world as it ought to be.[39]

Goodpaster and Respect for Self-Sustaining Organization

Citing such examples as endangered species, animal consumption and experimentation, abortion, and the uses of medical technology in general as motivating renewed attention, Kenneth Goodpaster also revisits the question of the relevant criterion for moral considerability in his article "On Being Morally Considerable." Such issues pose fundamental challenges to the "breadth of the moral enterprise" as conceived from the "humanistic perspective" of modern moral philosophy, leaving the field largely silent on the question of moral considerability in nonparadigmatic (nonperson) cases; and he believes that extension beyond the traditional criterion "may provide both a meaningful ethical vision and the hope of a more adequate action guide for the long-term future."

Goodpaster's thinking is more closely aligned with Taylor's than with either Regan's or Singer's in the breadth of its expansion of the moral community. "Neither rationality nor the capacity to experience pleasure and pain seem to me necessary (even though they may be sufficient) conditions on moral considerability," he writes. Sentience, for example, is "ancillary to something more important": it is "an adaptive characteristic of living organisms that provides them with a better capacity to anticipate, and so avoid, threats to life." And there is no special difficulty, he argues, in acknowledging and representing the interests of such nonsentient organisms as plants (their need for water and light in order to maintain themselves and grow, for example). It is life itself that is the ultimate value, and so he concludes that "nothing short of the condition of *being alive* seems to me to be a plausible and nonarbitrary condition." Noting that the defining characteristic of life is something to the effect of "self-sustaining organization and integration in the face of pressures toward high entropy," Goodpaster posits that "this criterion, if taken seriously, could admit of application to entities and systems of entities heretofore unimagined as claimants on our moral attention (such as the biosystem itself)." Ecological systems have feedback processes and energy pathways that maintain order (low entropy) and certainly they can be harmed and benefited, leading Goodpaster to ask, "Why should the universe of moral considerability map neatly onto our medium-sized [and, we might add, individualistic] framework of organisms?"

Of course, acknowledging the considerability of all living things, potentially including systems as well as individual organisms, complicates the question of trade-offs and conflict resolution. Although he does not develop a systematic prioritizing mechanism as Taylor does, Goodpaster does offer a distinction between the moral considerability of an organism and its "moral significance": the latter might legitimately vary within the set of moral subjects, as there is no particular reason to insist that the criterion determining standing be the same as the criterion whereby we prioritize or "weight" competing claims. Thus dogs, trees, and people all deserve moral consideration, they are all members of the moral community, but they may have different degrees of moral significance that legitimately justify trading off the interests of one for those of another. As Goodpaster concludes of the implications of his "life principle":

> We must eat, and usually this involves killing (though not always). We must have knowledge, and sometimes this involves experimentation with living things and killing (though not always). We must protect ourselves from predation and disease, and sometimes this involves killing (though not always). . . . The moral consideration due to all living things asks . . . for sensitivity and awareness, not for suicide (psychic or otherwise). But it is not vacuous, in that it does provide a certain ceteris paribus [other things being equal] encouragement in the direction of nutritional, scientific, and medical practices of a genuinely life-respecting sort.[40]

Ecocentrism: Leopold and Callicott

Goodpaster's interpretation of the criterion of "being alive" to include not only individual animals and plants but potentially also the ecosystems they constitute highlights a central point of disagreement within the extensionist project: whether environmental ethics must be grounded in the considerability of individual nonhuman organisms or whether it can depart from the human model of atomistic rights-holders to account for the considerability of communities or larger systems. Singer, Regan, and Taylor take the individualistic approach; they defend the moral considerability of individual nonhuman organisms and, in fact, explicitly deny the possibility of larger wholes such as species or communities being granted moral considerability. Singer, for example, writes that "species are not conscious entities and so do not have interests above and beyond the interests of the individual animals that are members of the species," and he thus has no argument for the preservation of species as valuable in and of themselves other than the assertion that "defending endangered species is, after all, defending individual animals as well."[41] Regan, arguing that "paradigmatic rights-holders are individuals," has coined the term "environmental fascism" for holistic approaches to environmental ethics that privilege species or communities as such over individual rights-holders.[42] And Taylor similarly writes that "the population has no good of its own, independently of the good of its members" and that "the good of a biotic community can only be realized in the good lives of its individual members."[43]

Just as interhuman ethical theory is increasingly being challenged to accommodate the rights of collectives such as corporations, racial groups, and the human species itself, so is the extensionist project being challenged from within on the grounds that its largely individualistic theorizing is incapable of expressing the values of populations, species, communities, and ecosystems as such. J. Baird Callicott, whose text *In Defense of the Land Ethic* offers perhaps the most comprehensive challenge to individualism in extensionism, presents an alternative holistic perspective, strongly rooted in the work of Leopold. Calling his approach "ecocentric," Callicott advocates "a shift in the locus of intrinsic value from individuals (whether individual human beings or individual [animals]) to terrestrial nature—the ecosystem—as a whole."[44]

According to Callicott, Leopold clearly calls for an enlargement of the moral community to encompass more than our own species, but his nonanthropocentrism is focused on ecological communities, such that the locus of value is not the individual organism but the larger systems they comprise and the natural processes that are fundamental to all life. Ecologist that he was, Leopold couched his extensionism in the context of a natural progression toward an ever-broader understanding of the concept of *community*: "All ethics so far evolved rest upon a single premise: that the individual is a member of a community of inter-dependent parts. . . . The land ethic simply enlarges the boundaries of the community to include soils, waters, plants, and animals. . . . The extension of ethics . . . is . . . an evolutionary possibility and an ecological necessity."

If we are now at the point of consciously extending the range of our ethical concern beyond the human species, this is the same dynamic that characterized the historical inclusion of slave-girls and more recently of racial minorities and of women, and that dynamic is one of recognizing these others as mutual members of a "community of interdependent parts." And as in these previous expansions, understanding "the land" as "a community to which we belong" rather than as a "commodity belonging to us" leads us from abusing it to treating it with "love and respect." This latest extension, however, may be especially difficult because "civilization has so cluttered [the] elemental man-earth relation with gadgets and middlemen that awareness of . . . our dependency on the soil-plant-animal-man food chain . . . is growing dim [and] we fancy that industry supports us, forgetting what supports industry." Available to us since the time of Darwin, though, is "new knowledge" that should encourage us to think of ourselves as "cog[s] in an ecological mechanism" and as "fellow-voyagers with other creatures in the odyssey of evolution"; this "sense of kinship with fellow-creatures" ought to inspire "a wish to live and let live; a sense of wonder over the magnitude and duration of the biotic enterprise." Part of our emerging consciousness of our place in the broader ecological and evolutionary context involves learning to "think like a mountain": to place our own short-sighted interests in the context of historical time and of complex ecological interactions. Corollary to this increasing awareness of our interdependence with the environment, then, is an enlargement of the boundaries of the moral community in accordance with an ecological understanding of community.[45]

The primary consequence of this awareness of interdependence and of the subsequent expansion of the moral community is a new way of understanding our place in the world. "In short," Leopold writes, "a land ethic changes the role of *Homo sapiens* from conqueror of the land-community to plain member and citizen of it. It implies respect for his fellow-members, and also respect for the community as such."[46] It is from this last phrase and the central statement of Leopold's "land ethic"—"A thing is right when it tends to preserve the integrity, stability, and beauty of the biotic community. It is wrong when it tends otherwise"[47]—that Callicott derives his holistic (versus individualistic) version of extensionism. "The good of the biotic community," he writes, "is the ultimate measure of the moral value, the rightness or wrongness of actions. . . . [and] the moral worth of individuals is relative, to be assessed in accordance with the particular relation of each to the collective entity which Leopold called 'land.'"[48] Thus, natural systems are the locus of considerability in Callicott's ecocentrism.

And here again our actions in the natural world are to be constrained but not to the exclusion of our need to fulfill our own interests as biological entities. Leopold writes that "a land ethic of course cannot prevent the alteration, management, and use of these 'resources,' but it does affirm their right to continued existence, and, at least in spots, their continued existence in a natural state."[49] So what are the specific implications of Callicott's development of Leopold's land ethic? "In every case," Callicott summarizes, "the effect upon ecological systems is the decisive factor in the determination of the ethical quality of actions . . . [and] differential moral value [is assigned] to the constitutive individuals relative to that standard."[50] Thus, an ecocentric environmental ethic might justify or even require deer hunting, for example, in regions where an otherwise unregulated population might threaten the local biotic community, and it might similarly condone trapping beavers or removing their dams in order to protect the river community from the damaging effects of siltation. Alternately, predators and rare animals "should be nurtured and preserved as critically important members of the biotic communities to which they are native." The land ethic grants "inanimate entities such as oceans and lakes, mountains, forest, and wetlands . . . a greater value

than individual animals"; and it recognizes the "greater claim to moral attention . . . [possessed by] animals of those species, which, like the honey
bee, function in ways critically important to the economy of nature [in comparison with] psychologically more complex and sensitive [animals], say,
rabbits and voles, which seem to be plentiful, globally distributed, repro-
ductively efficient."[51]

Ecocentrism thus adds to extensionism's challenge to a narrowly delimited (anthropocentric) moral community an ecologically informed critique
of strictly individualistic conceptualizations of value. While it is undoubtedly more complex and probably of less-established philosophical lineage,
this approach is not without similar parallels in interhuman ethics:

> It is not uncommon in historical moral theory . . . to find that in addition to
> those peculiar responsibilities we have in relation both to ourselves and to other
> persons severally, we also have a duty to behave in ways that do not harm the
> fabric of society per se. The land ethic, in similar fashion, calls our attention to
> the recently discovered integrity . . . of the biota and posits duties binding upon
> moral agents in relation to that whole. Whatever the strictly formal logical con-
> nections between the concept of a social community and moral responsibility,
> there appears to be a strong psychological bond between that idea and con-
> science. Hence, the representation of the natural environment as, in Leopold's
> terms, "one humming community" . . . brings into play . . . those stirrings of
> conscience which we feel in relation to delicately complex, functioning social
> and organic systems.[52]

Conclusion

To summarize, then, what are the primary elements that characterize the
first of our conceptual lenses? We have examined this question largely by
tracing the evolution of one of the central concepts in moral philosophy, that
of the moral point of view, which, in microcosm, embodies the spirit of this
perspective on ethics. And we have seen that extensionist environmental
ethics is modeled in many ways on the assumptions and perspectives that lie
at the very heart of the tradition of rationalism. We conclude this chapter
with a brief comparison of extensionism and the larger philosophical tradition of which it is a part, in order to review and firmly establish the key elements as they have evolved within this tradition (both the extent to which
the primary characteristics of rationalism are imported into environmental
ethics theorizing and the nature of their sometimes altered appearance in
this newer body of theory). The following seven elements, of course, over-

lap significantly and are not strictly distinct, and they are not necessarily exhaustive; readers are encouraged to identify other such features that define the crucial substance of this framework. These are among the primary characteristics we will be looking for as our imaginary extensionist ethicist reconstructs the story of the gray whale rescue from within this framework in the next chapter.

The Primacy of Reason

First, this approach hinges crucially on the primacy of reason. As humans, we are understood to be, above all, rational beings, and it is in the exercise of our rational faculties that we epitomize the human ideal. Further, ethical dilemmas—like the very structure and function of the world at large—are understood to be rational in nature, resolvable through the use of logic and reason. In our capacity as moral agents, therefore, we are enjoined to rely on the principles of mathematics, of logic, and of sound reasoning from general principles to particular conclusions in making ethical decisions. We are given rigorous deliberation procedures that both grow out of and reinforce our understanding of the world and ourselves in distinctly rational terms: adopting the moral point of view means, first and foremost, engaging our powers of reason. Moral failures are thus essentially the product of errors in reasoning or, in other words, of incomplete information, of allowing one's judgment to be swayed by emotion, of insufficient clarity with regard to moral concepts, or of inadequate reasoning through consequences.

The primacy of reason is a dominant characteristic of extensionist environmental ethics as well. Often the very impetus toward extension of ethical theory is recognition of the inconsistency or arbitrariness inherent in its person-limited realm of application; anthropocentrism is thus challenged on the grounds of its irrationality. And regardless of how their critics receive the ethical positions themselves, the works of Singer, Regan, and Taylor are each lauded as testaments to the role of rationality in philosophical discourse. *The Case for Animal Rights* is undoubtedly one of the most rigorously defended arguments in the entire philosophical canon, reflecting the author's "sustained commitment to rational inquiry" and to making "a reasoned case."[53] Taylor similarly describes his theory as "an attempt to establish the rational grounds for a system of moral principles . . . which requires systematic examination of a broad range of problems" and notes that "establishing [its] rational acceptability [requires] . . . mak[ing] every effort to achieve objectivity of judgment and . . . philosophical reflection."[54]

The Corollary Devaluation of the Emotions

Related to the primacy of reason is the corollary devaluation of the emotions or, in general, of the nonrational aspects of being human. Personal affections, friendship, and love are seen as obstacles to be overcome in achieving the necessary impartiality; they are illegitimate influences that have little if any place in well-reasoned deliberation. We may act out of love or friendship or concern, but we may not consider such action to be "ethical" or to have any moral value; acting out of mere sentiment is not in the same category as action based on principled reflection. In general, subjective states of mind—affections, fears, desires, intentions—are less significant, less meaningful or real, than are the quantitative and measurable aspects of objective reality; and the emotions, specifically, are arbitrary and capricious compared with the consistent, cool, and purposeful operation of an informed and rational intellect.

Such is also the case in extensionist theorizing. Each of our extensionist authors goes to great lengths to dissociate his argument from the sentimental overtones of the animal and environmental movements and thus attests to the negative value of emotions in environmental ethics theorizing. Singer, for example, goes out of his way to tell us, in his Preface, that he doesn't "love animals" or own pets, and he dismisses with some condescension the common assumption that only "animal-lovers" would be interested in the question of the ethical treatment of animals. "The portrayal of those who protest against cruelty to animals as sentimental, emotional 'animal-lovers,'" he writes, "has had the effect of excluding the entire issue of our treatment of nonhumans from serious political and moral discussion," and thus he elevates his work above such charges by deliberately making "no sentimental appeals for sympathy." And not only are associations with sentiment responsible for downgrading the seriousness with which his topic is received, but emotional appeals are understood to be less effective to begin with in actually changing behavior: "kind feelings [are less] universal and . . . compelling in [their] appeal [than is reason] . . . and [although] everyone is at least nominally prepared to listen to reason, . . . an appeal to sympathy and good-heartedness alone will [not] convince most people of the wrongness of speciesism."[55] And, echoing Singer's argument about effectiveness, Regan states that "kindness" motivated by sympathy, affection, concern, or even love "simply will not do the job of grounding our positive duties to animals."[56] Nonanthropocentrism is understood to be a matter of rational principle, not sentiment.

A Worldview Based on Hierarchical Dualism

This dichotomy between reason and emotion is one example of the broader tendency of the tradition of rationalism to build on the hierarchical dualisms of Platonic and Cartesian rationalism. Thus, reason and emotion are mutually exclusive opposites, and reason is the more highly valued member of the pair. In the same fashion, the distinctly human subject is understood to be fundamentally separate from and above the world of nonrational objects that, subsequently, exists for his instrumental purposes; the world of human culture and human progress is thus superior to the natural world, the world of "things." Similarly, the moral point of view maintains this same type of gulf between "self" and "other," between the moral agent and the object of his deliberations; it enjoins a distancing of the two and a dispassionate consideration of the "other."

The role of hierarchical dualisms in extensionist thought is a bit more complex. The strict and value-laden dichotomy between the human and nonhuman realms in general, which is characteristic of Cartesian rationalism, is a large part of what environmental ethics challenges; thus that particular dualism can have no place in nonanthropocentric ethical theorizing. The tendency to categorize the world in dualistic terms and then to prioritize one aspect over the other, however, is retained in the concept of the expanded moral community; this is seen as a necessary component of any ethical theorizing (after all, how could value possibly have any real meaning if everything had it?) and is taken to be unproblematic insofar as it is no longer arbitrary as in its original Cartesian formulation. The goal is not to eliminate hierarchical dualisms from our theorizing but rather to take more care with respect to where we draw the line of moral considerability, accepting only that line drawing that can survive the demands of consistency and of nonanthropocentrism. The entire approach of broadening the scope of moral considerability assumes that there will still be a fundamental distinction between those who are "in" and those who are "out" and consequently a significant basis for justifying differential treatment (unless, of course, the moral community is taken to include everything that is, a position that few if any extensionist authors take seriously). Hierarchical dualism persists in Singer's distinction between sentient and nonsentient beings, in Regan's discrimination among animals who are and are not subjects-of-a-life, in Taylor and Goodpaster's demarcation between living and nonliving, and in Callicott's ecocentrism (where the hierarchical dualism takes the slightly different form of distinguishing and privileging morally relevant ecological "wholes" such as species, communities, and ecosystems over individuals—human or non-human—whose value is merely instrumental with respect to the good of the whole). Finally, it should be clear that each of these environmental ethics theories incorporates adoption of the moral point of view in ethical deliberation and thus maintains this tradition's characteristic separation between the moral agent and the object of his deliberation; expanded hedonistic calculus, balancing animal and human rights, and applying the rules for resolving conflict between respect for persons and respect for nature all require dispassionate, objective, distanced consideration of human and nonhuman others and thus depend crucially on the dualism of self and other.

A High Level of Abstraction

These dualisms and the separation they maintain between the categories are one example of the high level of abstraction characteristic of the rationalist tradition. We are to adopt the Cartesian worldview, to understand the world and ourselves in dualistic terms, independently of whatever our actual experience may tell us. The moral point of view itself is an abstract perspective. Not only is it a mental construct, but it also enjoins a decision procedure that moves from the general to the particular (from the abstract principle that "Lying is wrong" to the particular conclusion that "I should not lie to my mother about being out late last night"); attention to the particular details of the situation at hand, to its concrete and unique reality, is less significant than characterizing it in terms of the universal constants it shares with other abstractly characterized situations. Similarly, personal involvement in the situation presenting the ethical dilemma, or empirical experience of any sort, is neither helpful nor necessary in ethical decision-making because legitimate answers to the dilemma are to be found by appeal to general principles that would be relevant to any moral agent in any similar circumstance.

The injunction to adopt the moral point of view in our environmental ethics theorizing is the defining example of the high level of abstraction extensionism shares with the rationalist tradition. The pattern of reasoning from general principles to particular situations, the characterization of moral conflict in universal terms without necessary reference to concrete details, and the understanding of principles as applying to moral agents interacting with the natural world *as* moral agents rather than as participants in any given dilemma are all retained in the works of the extensionists. Although *Animal Liberation* presents in detail the actual treatment of animals in the factory farming and animal research industries, the argument for eliminating these practices is itself abstract (although perhaps less so than is Regan's) in that the principle of equal consideration of interests does not emerge from awareness of the harsh treatment of animals but rather is assumed a priori and applied to the situations as they arise. Regan's theory of animal rights, deontological rather than utilitarian in nature, is even further removed from any need to consider the details of particular instances of animal exploitation; the purpose of any given animal experiment or the various contexts in which animals might be raised for food are irrelevant, and thus all such issues are cast simply and abstractly as violations of animals' rights not to be treated as mere resources.

The Application of Universal Principles

The abstract quality of rationalist tradition theorizing also emerges in the pattern of applying general rules or universal principles to the particular situation at hand. This reliance on universal principles denotes the rule-governed nature of morality. These principles are understood to be universal in the sense that they dictate the moral agent's response, not with regard to his individuality or to the unique elements of the situation, but rather *as* a moral agent facing the abstractly defined (without reference to the particular situation at hand) choice between lying or not lying (for example). Ethical deliberation need not be sensitive to context but rather must begin with consideration of the general principle or principles at stake. This is not to suggest, of course, that universal principles are absolute or cannot be overridden by ethical considerations that might come into play variously from one situation to the next, thus the need for a hierarchy among principles.

The retention of this characteristic within extensionism is demonstrated most clearly in Taylor's *Respect for Nature*. Attesting to the rule-governed nature of morality, Taylor's ethical system hinges on the careful working out of "rules and standards [that] constitute a system of ordered principles;" his four rules of duty "tell us what general kinds of actions we are morally required to perform or refrain from performing," and, similarly, the five priority principles adjudicate conflicts between the duties derived from the principle of respect for nature and those derived from the principle of respect for persons. In adopting the biocentric outlook and accepting these rules and principles as valid norms that are morally binding on our behavior, "we at the same time believe that they are morally binding upon all other moral agents;" thus they are understood to be universal or applicable by moral agents as such, regardless of the unique situation in which the moral agent may find himself. The rules of nonmaleficence, noninterference, and fidelity, to use the rules defining our negative duties in this example, "lay down a prohibition upon all actions of a certain type, prescribing that *any* moral agent refrain from *any* action of that type."[57] Thus, the fundamental principle of respect for inherent value (whether of humans or nonhumans) is understood to apply to all moral agents as such, and the determination of the appropriate course of action in any particular situation proceeds by identifying the relevant rules and then working out what specific response those general rules call for in that situation.

The Importance of Impartiality in Decision-Making

Further testifying to the universality of principles, to the autonomy of the moral agent, and to the irrelevance of emotional attachments is the importance of impartiality in ethical decision-making. Human beings are to be respected, or our interests taken into account, simply by virtue of being rational or sentient beings affected in some way by the dilemma at hand. The fact that my friend is involved in the situation and that my concern for her outweighs my concern for the similarly situated strangers involved can have no bearing in my deliberation; her rights or interests can carry no more weight than those of any other parties (recall the utilitarian injunction: "Each to count for one and none for more than one"). Similarly, my own interests, or those of any group which I may be a member of or otherwise have sympathy with, cannot be given a prior or heavier weighting and thus allowed to sway or bias the ultimate decision. In many ways it is impartiality in ethical deliberation that the moral point of view is explicitly conceived to address: this objective and dispassionate perspective is immune to the inappropriate biases inevitably associated with a more personal perspective.

The tradition of rationalism's requirement of impartiality in ethical deliberation lies at the core of the extensionist critique. Anthropocentrism is, by definition, perhaps the broadest form of bias or partiality: bias in favor of one's own species at the expense of other species. Extensionism, therefore, not only retains the emphasis on impartiality but also enjoins us to practice it on various levels. We must be impartial between our own interests as individuals and the interests of individual and collective nonhumans (such that favoring our own interest in eating meat over a pig's interest in freedom and life or favoring our own interest in having a new mall over the integrity of the ecosystem that would consequently be disrupted both constitute unjustifiable bias), between the interests of our species and those of other species (such that automatically eliminating the smallpox virus in the name of human health constitutes partiality), and between the interests of the various nonhumans who are affected by our actions (such that we cannot justifiably favor "charismatic megavertebrates"—those large mammals like chimps and koala bears that we find it so easy to sympathize with—over ugly or otherwise innocuous species such as cockroaches and snakes). The characterization of Singer's theory in terms of "animal liberation," like the use of the term "speciesism," draws the parallel between unjustifiable partiality for one's own species and the similar historical bias toward one's own gender (sexism), race (racism), or class (classism); Singer insists that the interests of any being that suffers (regardless of his species) are to be counted equally with the like interests of any other such being and thus ensures the impartial consideration of human and nonhuman interests. Taylor similarly condemns our tendency to "feel sympathy for the prey and antipathy for the predator" and to "favor certain species over others and to want to intervene in behalf of [our] favorites." "To respect nature," Taylor summarizes, "is to be willing to take the standpoint of each organism, no matter what its species, . . . and to [refuse to] count the good of some as having greater value than that of others."[58]

An Understanding of Morality as Adjudicating Conflict and Ensuring Justice and Fairness

Finally, the importance of impartiality in this tradition is due in part to the understanding of morality as adjudicating conflict and ensuring justice and fairness: ethical dilemmas are most often cast in terms of competing rights claims or incompatible interests, and giving each person the consideration he or she is owed *as a person,* independent of any personal stake or attachment the moral agent might feel, is the essence of a just resolution of such conflict. The rationalist model of human social interaction is an atomistic, competitive one, which posits individual, autonomous agents acting largely out of self-interest and thus clashing with one another constantly in a world of limited potential for interest fulfillment; ethical dilemmas are thus interpreted as interpersonal conflicts, and it is morality that promotes the smooth running of society by resolving these conflicts through the rational weighing of interests, the balancing of rights, and the application of universal principles of justice.

Extensionism replicates the rationalist tradition's understanding of morality as adjudicating conflict and ensuring justice and fairness. An expanded moral community necessarily entails conflict among the increased number of beings granted moral considerability. Within its largely atomistic and competitive framework, individual human and nonhuman organisms are understood to be fundamentally in conflict, their rights often incompatible and their interests mutually exclusive. Rights, in fact, are understood to exist solely for the purpose of protecting one individual from the incursions of others, and this is no less true of Regan's animal rights theory than of Kant's human rights perspective; "in issuing its condemnation of established cultural practices," Regan writes, "the rights view [be it pertaining to the human or the animal realm] . . . is simply projustice."[59] Justice, or fairness of treatment and the balancing of benefits and burdens, is, in fact, the core concern of the extensionist project. Just as it is unjust to subordinate the interests of African Americans to the interests of Caucasian Americans simply on the grounds of skin color, so too (and this is the nonanthropocentric thesis) is it unjust to restrict consideration of rights or interests to one's own species on the basis of irrelevant differences. And it is unjust or unfair to automatically resolve the conflicts between ourselves and nonhumans in our own favor. Thus, both Singer's call to weigh conflicting interests and Taylor's discussion of competing claims and the principles for prioritizing them assume conflict among the members of an expanded moral community and attest to the adjudicatory, justice-ensuring function of morality.

I know you will try to save them. They have a right to live just like us.

In one view, at least, the tale of the gray whales of Point Barrow speaks volumes about misplaced priorities and the goofiness of man.

Chapter 3 Whale Rescue Story I

We are now well prepared to take a close look at our case study in light of the conceptual framework sketched in the preceding chapter. Our project here will be to tell the story of the gray whale rescue from the perspective of the tradition of rationalism, specifically from that of extensionism. In doing so, we will deepen our understanding of the event itself, beginning to find answers to the question of *why* it happened, and we will also come to better understand the perspective embodied in this dominant philosophical tradition. Seeing it in action, so to speak, will sharpen our grasp of the characteristics that uniquely define the first of our conceptual lenses, and we will begin to appreciate the powerful role of such lenses in shaping our interpretation of events such as the whale rescue.

As we noted in the Introduction, this project entails emphasizing certain elements as key to our understanding of the event while omitting others as irrelevant or meaningless. We might expect, for example, that this version of the story will focus on the rationality or irrationality of the decision to attempt the rescue and that it might cast supporters and detractors as representing opposite poles of the anthropocentric/nonanthropocentric and the individual/system spectrums. Similarly, we might expect the issue to be characterized abstractly and in terms of conflicting interests—humans versus whales, for example—and we might anticipate a central role for concepts such as duty and principle.

Recalling these and other characteristics of the perspective embodied in the rationalist tradition and its offspring extensionism, let us turn now to the four central questions to be asked of the case from the perspective of each of our three lenses. It may be helpful to approach this chapter (and its counterparts later in the text) as if you are listening to an ethicist who works within this tradition explain the various facets of the event from the perspective afforded by this tradition; you might imagine that you are posing each of our four questions to this imaginary ethicist and then listening to his or her responses. It is important to note, of course, that this interpretation of the event, drawn from the comments of participants and observers, is representative of this perspective but certainly not exhaustive of the possibilities for making sense of the event even within this one lens.

Why was the rescue effort the appropriate course of action?

(1) It was grounded in respect for inherent value.
"It was our obligation to save the whales if we could." "[Once] we knew about it and we had options to get them out, [it became] a moral thing." And as a "moral thing" it transcended economic concerns. Thus, neither the costs nor other potential uses for the funds spent on the rescue were relevant in the response to the whales: one participant noted of the critical response to their having spent the money on the whales rather than on human needs that "that wasn't the debate," and another noted that "there was nobody . . . who said, 'Wait a minute guys, let's put this down on paper and let's do some economic assessments of what we're getting ourselves involved in.'" The lives of the three whales were not something "that you can put a price on." In deciding to attempt the rescue operation, no one set a monetary value on the whales' lives with the intent of aborting the effort once the costs exceeded that value; once the mounting costs did become an issue it was still not a matter of putting a price on their lives but rather of reasonable allocation of financial resources among competing ethical obligations.

"Knowing that other creatures have an inherent worth just like we do, . . . we have this moral obligation to help out our fellow species." It is not the case that only human beings matter, that we alone are inherently valuable. Opposition to the rescue on the grounds that it diverted resources from pressing human needs simply attests to the outdated but "prevalent vanity in our way of thinking that . . . we are the most important living organisms on earth" and that "our problems and our welfare are somehow more important than any other creatures." "Actually, we are just another link in the chain, humbling as that fact may be," and this understanding of our place in the world leads us quite naturally to see the same inherent value in animals such as whales that we have always seen in ourselves. We are not fundamentally superior to all other forms of life, and they are not simply resources that we can use when we like and otherwise ignore: "They are not things. Anything that lives and breathes is not a thing; it is a being in and of itself."

A basic respect for life justified the rescue: "They were alive. There's a respect for life hopefully all people would have. . . . The life force of the whales, . . . that they were living creatures, sentient beings, . . . was probably the bottom line." It was a universal obligation, imposed on us as moral agents regardless of our proximity to the situation or our personal feelings about whales, so it was to be expected that people all over the world would share this sense of obligation: "That such an adventure could affect and involve so many people from different nations and walks of life is a testament to the majority of humankind's . . . sense of right." Several participants took care to clarify that their support of the rescue was a matter of principle and not merely of sentimental attachment to animals as many detractors charged, claiming "I concurred . . . that it was the right thing to do and I'm not really a great animal fan." Another summarized "the great principle involved": "These creatures share our environment and we should reach out to help them in times of great emergency." Our increased awareness of ourselves and other creatures as sharing in inherent value and thus as being equally subject to ethical consideration justified the rescue effort as the "right thing to do."

(2) The whales' suffering was morally relevant.[1]
For almost everyone involved, leaving the whales to die painfully was never an option. As one of the local officials stated, "These whales are suffering and need our help; we feel a duty to do what we can." "It was either put them out of their misery or do something [to help them] because they were suffering," and, in light of "their obvious and growing misery," a Barrow hunter voiced one of the earliest suggestions for dealing with the whales' situation: "I suggest mercy killing. This is awful." "Just like you might pull the plug on your father or your mother at some point" to end the pain of a prolonged and inevitable death, so too would simply shooting the whales to end their suffer-

ing have been a legitimate, perhaps even an obligatory, response to the situation. This alternative, however, was the second-best response, second to ending their suffering by setting them free from the ice; and thus the option of mercy killing was kept in reserve but was held secondary to the option of attempting a rescue. In the words of one participant, "If no one is able to help them, then we want to stop them from suffering."

(3) We had an obligation to balance the scales of justice.
Our obligation was also grounded in the need "to balance the equation a little bit" with respect to our historical mistreatment of whales. Gray whales have been hunted almost to extinction twice, and they were still listed as endangered at the time of the rescue. Most of the news coverage of the event described them as "rare," giving support to the commonly expressed viewpoint that we "humans owed those whales the chance for survival [since] after all, it was [we] who caused them and other forms of wildlife to be placed on the endangered species list." We had an obligation to help balance the scales of justice, offsetting the excessive burdens we have imposed on them for our benefit by here assuming a burden for their benefit.

(4) Much of the opposition was irrational.
Proper ethical deliberation requires consistency within an expanded moral community, and arguments that we should not interfere with the course of nature were terribly and self-servingly inconsistent. As many supporters asked, "would [detractors] . . . apply that philosophy to all inhabitants of this planet"—"without prejudice"—and thus allow the "weak or foolish" among human society to die naturally when they could be saved? Quite the contrary, many people saw the plight of the whales as the equivalent of the previous year's rescue of Jessica McClure, the little girl who fell down a well in Texas, and they no more considered leaving the "foolish" whales to their fate than they did leaving the "foolish" child to hers. As one reporter aptly expressed the parallel, "The icy waters off Alaska were their abandoned well. They were brave little toddlers. We couldn't resist." What we could not, and should not, resist is the duty to save life, in either case, regardless of their "foolishness" in having gotten into their respective predicaments and despite the natural selectionist argument in favor of allowing such "foolish" genes to be eliminated from the population. "Every whale is worth the effort" of saving just as every child is, and it is a testament to our ethical progress that "there are some people who see us all [—baby girls and whales—] as these sentient beings that . . . all matter." And, of course, it was most illogical for

the detractors to oppose the rescue effort on the grounds that it diverted funds from pressing human needs, because the funds would not have been spent on human beings instead if the rescue had not occurred.

What was the relevance of their being *whales* rather than another type of animal?

"Just the fact that they were whales" was indeed a significant part of the decision to attempt a rescue and an important impetus to the widespread interest in the event. We can explain the comment of one observer that "it definitely seems to be a species that there's something special about for a lot of humans" with reference to two specific facts about whales: they have great symbolic value and they are in many ways similar to our own species. In both cases, the significance of the species (actually, their order, *Cetacea*) of the trapped animals attests to the generality and abstraction of proper ethical deliberation; it was not the individuality or uniqueness of these three animals but rather their being whales that (legitimately) played a role in the decision-making process. Interest was sparked in the rescue effort because it was a chance to "save the whales."

(1) Whales have symbolic value.
"They were three whales but they did stand for something much more than that; . . . there was something more than just three whales." One observer explicitly noted, "I don't think the two whales were the point; they were a vehicle; . . . they were symbols." Whales are "icons of the environmental movement," and participants carried this symbolism with them into their encounter with these three whales. The whales were "symbolic of the continuum of life," and their "plight was a . . . metaphor for the dramatic question of life's continuity in general on this precious planet." With whales in general "tagged as the tragic symbol of man's overexploitation of nature," these three whales came "to symbolize survival": of all threatened species of whales, of the environment in general, and of humanity itself. Whales in general are a "national treasure," and thus these three whales in particular were treasured.

(2) The similarity between whales and humans renders them obvious members of the moral community.
As whales these three animals were "in many ways so much like man"; and those nonhumans that are most like ourselves are the ones most obviously

"deserving" of the same type of moral consideration we acknowledge in one another. "If they hadn't been so human . . . they probably wouldn't have been fought as hard for." "Whales are, like man, warmblooded, air-breathing mammals" "[that] sing and all," and "we were inclined to help them [because] . . . they are . . . mammals [and because] they have a high degree of intelligence." In the words of one commentator, "What sticks in the memory is that heavy breathing. Breathing is something we share with the whales. They are mammals like us, and that makes for a certain kind of kinship." In this light, the oft-noted anthropomorphic tone of much of the contemporary commentary makes sense: because the extension of moral considerability proceeds first to those animals sharing the characteristics we most value in ourselves, the rescue effort gained legitimacy as a moral obligation to the extent that the three whales and their plight were understood in human terms. Even detractors acknowledged the dynamic according to which we came to treat these whales much the same as we would treat other people: "There's this idea that they are as smart as we are, that they are no different than we are. . . . If [people] want to believe that, then they get into this next plane that its not very far between that whale and a person being caught in that hole." The fact that the trapped animals were whales was indeed significant to the instigation of the rescue effort and to the public's interest in their plight; this significance was best captured in the words of one participant: "I had the same response I would if I saw a person in trouble. [Even though they are not people?] Yeah, they are."

What was the basis for opposition to the rescue effort?

(1) It was irrational.
One commentator summed up this first position, asking, "Should we applaud the dubious motivation of virtually everyone involved in the spectacle? How could such an unreasonable undertaking give dignity to anyone involved in the affair?" The "dubious motivations" were the sentiment and emotion, the "collective caterwauling," that in many people's eyes drove the rescue effort. Seen in this light, the rescuers were "investing untold time and energy in reaction to the emotions of people who do not understand the natural world." The public's response to the whales was a "silly . . . entirely sentimental . . . irrational response—the way people react to kittens and puppies"; reporters, in fact, both fed and appealed to this response because "nobody was expected to think . . . once they got to Barrow" and because "[as] P. T. Barnum said, no one ever went broke underestimating the intelligence of the American people

. . . [and] readers are silly sometimes—absolutely [they were this time]." The sentimental public perceived the whales simply as "giant pets" and responded to their situation with an attitude of "Oh, aren't they cute," just as they would to a "kitten in [a] tree"; "there were a whole lot of [people] who didn't want to believe" that the youngest whale died during the rescue effort, people who irrationally held onto the belief that he "could hold his breath long enough to find [open water] . . . [and] that there was some chance that [he] might have escaped." Detractors took care to clarify, however, that their opposition was not similarly driven by emotion, as in the words of one participant: "my reactions to this . . . do not stem from a hatred of animals."

Suggesting that the rescue effort was characterized both by a predominance of emotion and by a lack of sound reasoning, one reporter wrote, "I was taught to value reason over emotion. It is a lesson many of those involved in Barrow seem to have forgotten." Although "certainly it is impossible to go out on the ice with the trapped whales and not feel empathy for them, maybe even shed a tear . . . man is an animal that is supposed to deal in more than just emotion[;] reason is what is supposed to set us apart from the other animals[,] and the cold harsh reality is that there is no reason to save these whales." Reason should have dictated that the effort was "unnecessary" to begin with and that its magnitude was "out of proportion with the problem" and with the actual chance of success, because "most of the informed people who were betting were betting against those whales ever seeing the light of day once they left." In the words of one detractor, "except for that emotional . . . kind of response that the least thoughtful and most impressionable people would have had throughout, anybody who thought about this at some point had to say, 'That's it, . . . it just [isn't] worth it.' . . . At least you hope that people are smart enough to reach that conclusion."

And, finally, from the perspective of the detractors, reason should have pointed out the inconsistency of the rescue effort. Not only do we "kill cats and dogs by the million and slaughter food animals in barbarous ways, [but surely] it is a bit silly to carry on at such lengths about two ice-bound whales when whales are still hunted with the blessings of Japan and Iceland for no reasonable purpose at all." Reason must judge as hypocritical Soviet participation in the rescue effort when that country has permission from the International Whaling Commission to kill approximately 170 gray whales annually on behalf of aboriginal populations; similarly with respect to the participation of the Inupiats: it "[does] not seem to make a whole lot of sense . . . [that] they're out there hunting these whales . . . and here they are saving three of them." Reason demands consistency, and because "the same people who seem so concerned about these whales generally turn a deaf ear to the concern for these creatures the rest of the time" (for example, "there was scarcely a ripple of reaction in the United States or elsewhere . . . when the Japanese announced plans to kill three hundred minke whales . . . a week or so earlier"), this rescue effort was irrational. It was similarly inconsistent to "root for the whales and ignore the fact that we dump poisons into the oceans with such zest that the water may become lifeless." From the detractors' perspective, the rescue effort resulted from failing to properly deliberate, from succumbing to inappropriately sentimental motivations, from allowing emotion and sympathy to subordinate reason; because of this, the rescue had no moral worth and, in fact, was morally problematic. "It defie[d] explanation for one very simple reason: it really doesn't make sense . . . by most rational measures."

(2) It was a conflict of interests between humans and animals that should not have been resolved in favor of the animals.
The comments of one observer highlight two other positions from which some people opposed the effort, suggesting the hierarchical ranking by which animals in general are valued less than humans and individual animals are valued less than the natural systems they comprise:

> It is arguable as an ethical proposition that the reaction to this and the amount of time, energy, and money spent on these whales says that we have lost a sense of the fundamental value of things: we should understand that at all times and in all places some things are more important than others. The lives of three California gray whales on that relative scale don't count for much. When we treat them as if they do then it seems that our moral or ethical system has failed. They weren't people and they weren't important animals in the larger scheme of things. How many ways can we rank this? Because compared [with] the other things that might have demanded the time and the attention of the people both reporting the story and the people focused on it, they weren't very important. By any imaginable scale that's true.

The second rationale for opposing the rescue effort hinges on defining the problem as a conflict between human interests and the interests of the whales: as one detractor summarized, "Whether or not one approves of the rescue effort really should have nothing to do with one's feelings about whales. The question is how one feels about people in need." In other words, it was deemed unjustifiable to spend resources on the rescue effort that could have been spent meeting human needs instead. "The money that was spent on the whales could have been used to feed the poor. Instead, we wasted money on three mammals. What about the human race?" In the words of one detractor, "The resources of our society are finite, we cannot do everything for everything. . . . We don't have to give anybody anything but we don't have to throw it away either. I love animals, but I love people more." And in the words of another, "I can think of a hell of a lot better uses for the millions of dollars that were wasted on this fiasco. . . . How many hungry people could have been fed on this totally futile effort? How many cold and homeless people could have been afforded shelter? . . . Just how stupid can we get?"

Although much of the opposition was blatantly anthropocentric, denying that the whales had *any* meaningful claim on us, some of the criticism came from a less extreme position, a position that recognized the whales' claim as valid but that nevertheless favored resolving the conflict in favor of overriding human interests. One observer pondered the issue in this light: "Should you spend millions of dollars to rescue whales that made a big mistake, or should you spend that money to help children, to help mothers, to help fathers, to help people that don't have work? . . . I don't think that balances very well, but I don't know." Another participant discussed the dilemma in these terms, recalling, "The first reaction we would get was, 'You guys are doing a marvelous thing.' The second reaction was, 'Why are you doing this? . . . How can you justify spending [money] on such an effort when people are freezing in the streets?' . . . And there is really no response to something like that." In this spirit, there would also be no adequate response to the question of why funds are spent on one human cause rather than another: in both cases, the competing claims are both seen as valid and the resolution of the conflict one way or another is simply a difficult, and sometimes indefensible, choice.

(3) Its emphasis on individual animals misplaced the value that should reside in the larger gray whale population or in the natural functioning of the ecosystem as a whole (ecocentric opposition).
Some people opposed the rescue on the grounds that it placed greater value on the whales as individuals than on the gray whale population as a whole or on ecosystem processes at large. As one detractor noted, "If all this effort and energy and concern was directed toward the population of whales as a whole, it wouldn't seem so hypocritical." Although "we have had a big environmental disaster with whales . . . we have just decimated them . . . [and] we have to be all for preserving and getting the numbers back where they should be," this is a duty to the population as a whole and it has no relevance

to the rescue of these three whales. "As individuals, [the three whales] really were not significant" because the gray whale population as a whole was thriving and "they weren't among the last of their breed." The value that they might have as individuals is thus dependent on their contribution to the population, and if this contribution is minimal then so is their individual significance. In the words of one detractor, "You don't completely ignore the individuals—you can't be completely cold-hearted—but, on the other hand, whether the population is doing well or not is sort of the bottom line"; responding to individuals is thus seen as a matter of emotion (not being "cold-hearted"), whereas reason dictates the health of the population as determinant of the appropriateness of intervention. Reason, casting the whales' plight as "natural," was understood to value the system and noninterference with natural processes whereas emotion, casting it as "pitiful," was seen to lead to misguided concern for and identification with the whales as individuals. While "you can't leave a little girl at the bottom of a well [because] . . . taking care of little girls is our collective human responsibility, protecting whales from the cruelties of 'Mother Nature, red in tooth and claw,' is not"; not only is it not our responsibility to "save" individual whales from natural processes, it is in fact problematic because it comes at the expense of the natural processes on which ecosystem integrity and the healthy evolution of populations depend. Saving these individual whales, deliberately returning their supposedly less-fit genes to the population, illegitimately subordinates the whole to the individual. "This event is totally insignificant to the population of these whales—insignificant to virtually anything except to these three individuals—and . . . nature doesn't care; critters die cruelly every day, people die cruelly every day. Nature doesn't care."

Further, "there are certainly far greater threats to the whale population than these three idiots being caught by the ice," and, in many people's eyes the rescue effort thus served to divert resources and attention from the more important population to the less important individuals: "The U.S. all too easily squanders its concern and resources on such individual rescue efforts, while programs that might benefit the whole species go begging." "The money spent on the rescue could substantially increase enforcement to prevent the illegal export of whale products," and the international attention would have been better focused on enforcing the whaling moratorium; "now that mankind is abusing the ecology of the remotest reaches of the globe, . . . many noble animals need saving. Heroic measures are in order, to save species not specimens." "You can save a lot of whales, or other living things, with the . . . money that [was] lavished to unstick this winsome duo."

What are the implications of this event for environmental issues in general?

(1) Environmental concerns are widely perceived to be in conflict with human concerns.

The nature of much of the opposition to the whale rescue suggests that environmental and human concerns are perceived to be mutually exclusive and even contradictory. As one commentator noted:

> On the front page of this newspaper, below the story about the whales, was a report that 5 million American children may suffer from chronic hunger. Below that story, at the bottom of the page, the Associated Press reported that 2 million people are on the edge of death from starvation in Sudan—the result of summer floods and years of war. Yet it is the whales, barnacled and bloody, that haunt, that capture our imaginations and our prayers. . . . What does this say about us? About our values?

The tendency to define the problem in this way suggests that when we bring the natural world into the realm of concern, ethical questions of where our duty lies are cast in competitive terms. Along the same lines, several editorial cartoons came out during the weeks of the rescue depicting the contrast between our response to whales in need and our seeming lack of concern for the suffering of our fellow humans: in one, homeless people wear whale costumes in an attempt to get help; in another, an American hostage held in Lebanon for three years similarly speculates that having a whale costume would expedite his release; and in a third, people dying of starvation in Sudan hear on the radio that "The world stood transfixed today by the heartrending plight of the gray whales." Virtually the only negative phone calls that came in to Barrow during the rescue effort were from people incredulous that money would be spent on the whales rather than on homeless, hungry, or out-of-work people. What all of these examples suggest is that many people see human issues and environmental issues as sharply opposed and judge it wrong to spend money on the needs of nonhumans while human needs remain unmet; and if this is indeed the case, then extensionism faces an uphill battle in convincing people to give a fair hearing to the rights and interests of nonhumans.

(2) Anthropocentrism is still the dominant mindset guiding our interactions with nonhumans.

Similarly, both opposition to and support of the rescue effort revealed the predominance of an anthropocentric mindset. Referring to the relative lack of opposition to the rescue of Jessica McClure, one observer attested to the partiality for our own species that may well be extensionism's most formidable obstacle: "It's a lot harder to get over the bar when you're talking about whales than when you're talking about a child. At some level we've got to want to protect our species in some structured and rational way as well as just instinctively." It was difficult for many of the participants and observers, most of whom "just don't equate whales with people," to even conceive of ethical obligations to nonhumans as a legitimate motive behind the rescue attempt; as one observer commented dismissively, "I suppose there were people who have such reverence for God's creatures that they feel the same about, well, everything from rocks on up—that anything that's more sentient than a rock has got exactly the same value as Albert Einstein—but I don't think that number is very large." And, interestingly, even the people who supported the rescue effort tended to justify it in anthropocentric terms, defending it as the right course of action with reference to human interests, not in terms of the whales' interests; the rescue was "good" or "right" because of the international good will it generated, because it brought ideological enemies (industry and environmentalists, for example) together for a common cause, because it offered good training experience for the military, or because it bought the whaling community of Barrow time and good publicity in their campaign to continue bowhead hunting.

(3) Sentiment is a powerful but problematic motivator for environmental concern.

Third, "for the long haul, environmentalists must respect rationality and hope to inform the public. Cashing in on sentiment to save three whales . . . was not sensible." Although from the perspective of many extensionists there were certainly legitimate, principled, reasonable grounds to justify the rescue effort, it is probably true that these were not generally the grounds appealed to in the public debate, and the environmental community does itself and its cause a disservice by either accepting or fostering the idea that environmentally conscious actions are a matter of sentiment rather than of reason and duty. To the extent that this was "just a sentimental identification, not with the whales but [with] their helplessness," the cause of environmentalism and specifically of environmental ethics was not advanced: "what is needed is an identification with whales and with all creatures that is not sentimental but is based upon a much deeper respect and reverence for life." Sentimentalizing the whales' situation, which included humanizing them for greater "pull" on our heartstrings, reinforces the public's bias toward certain types of animals, with the result that "most people only want to save Bambi-ized animals, like whales or baby harp seals. They're not interested in the sort of wildlife that isn't made into stuffed animals, much less in the long-term task of preserving natural habitats." Many people would even have approved killing any polar bears that threatened "their" whales even though reason dictates that "you don't shoot one endangered species to save another." And, of course, from the ecocentric perspective sentiment and emotion are additionally problematic in that they most often focus on individuals and thus support the displacing of value from the population or ecological system. One participant concluded that "the emotional reaction [demonstrated here] can sometimes be very misplaced . . . [and] much more detrimental than good" and asked "What happens to the environment if we . . . allow emotion to overpower reason in all our dealings with nature?"

The path to freedom. The series of chainsaw-cut airholes leading from the whales' original position four and a half miles across the ice toward open leads in the Chukchi Sea.

Bone, Bonnet, and Crossbeak share the airholes, regularly surfacing to blow and then diving beneath the ice.

Workers use chainsaws to cut rectangular holes in the ice and then push the sectioned slabs under the icesheet with harpoons. The whales surface eagerly in each new hole, but the water begins to refreeze quickly.

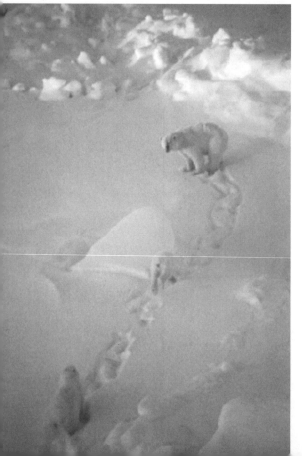

Polar bears, known to prey on ice-entrapped whales, are sighted in the vicinity. Workers begin carrying guns to the rescue site to ward them off if they approach the grays.

Part II Philosophical Tradition II

It's been almost seven years now since I first met them—Bone, Crossbeak, and Bonnet—but you know, "it would be hard for them not to still be with me." I've always wanted to go out to California and see the gray whales. "I always feel like Bonnet and Crossbeak (and, who knows, maybe Bone as well) are out there." And, of course, I especially think about them in October, the month they went through that terrifying ordeal up at Barrow, the month when I first became friends with three whales.

Some people may think it's silly or overly sentimental to talk about animals as friends, but that's what they are to me. "They will always be a part of me." "Whales are so easy to get emotionally attached to": they are enough like us that we can empathize with them, but they are also so different and mysterious that they can capture our imaginations. "People generally do feel a more intimate connection with whales than they do with other wildlife."

But with these three whales in particular, it was almost inevitable that we would develop relationships with them. "We all kind of started identifying with them after we were around them for a few days." "Out there on the ice, day after day . . . looking into their eyes . . . talking to them . . ."; "you really started to feel a kind of bond with them." Oh, and touching them! That was "the neatest portion of the whole thing," one of the most incredible experiences of my life. "Everyone went out and touched one of the whales, and they came away different after that." "That connection with actually touching a whale was very moving: then you began to root that they would succeed; you didn't want to see these guys drown, especially after you see them and touch them and hear them blow." It might be illogical and emotional, but you couldn't just "watch the hole freeze over and let them die when there was a way of keeping them alive."

And I think the same thing happened with people all over the country and all around the world: "you could sit in your living room and participate in the thing." "Television brought the whales right up close where you could look them in the eye," just like we could out there on the ice. "People actually saw the whales sticking their heads out of a hole in the ice and crying for help," and so they were just as interested in them and committed to helping them as we were. "When Bone showed signs of pneumonia, it was as though someone they knew had gotten sick." When he disappeared under the ice, "the nation mourned."

And the Inupiat Eskimos "have always had a special relationship with the whales; they've lived with and observed them for many years, and they really have a strong feeling of identity with them." "A bond that perhaps only hunters and hunted understand" made them work harder than anyone to free these three.

It seems pretty simple, despite all those writers going on and on about why in the world everybody responded to these whales the way we did. "Most humans just have compassion for something like that"; "we have a care instinct that responds to hurt and need," especially when we can focus our attention on particular individuals, be they whales or children or whatever. "Millions who remain unmoved by a generalized Save the Whales campaign were genuinely concerned about these specific whales, just as millions who seem indifferent to the Children's Defense Fund's urging on behalf of poverty-stricken and hungry children were willing to do whatever was necessary to rescue Jessica McClure from that Texas well."

So the tough part wasn't deciding to help the whales. The hard thing was knowing what would really help them. Of course we wanted to put an end to their suffering, but as the effort dragged on and on and we seemed to make so little progress, "it was frustrating to have the feeling that maybe we were just prolonging their misery and there was no way to save them and maybe it wasn't the right thing to do." But the scientists and the Eskimos paid very close attention to their physical condition and to their mood, and they did seem strong enough to have a real chance at survival. As long as there was a chance that we were actually helping them and not just prolonging their misery, we had to give them that chance.

Anyway, I am so glad we decided to try. "Not to help out some of our fellow beings when we had the opportunity would have shown that we really have turned into some pretty callous human beings." And I think it changed us all for the better: that kind of close, personal experience with animals helps us to feel a little less cut off from nature, maybe a little more aware of the interdependence between humans and nonhumans. Our experience with these whales helped us to see ourselves in a different light: "for a million dollars the world had the chance to get a little connection with another species for a while, and that's OK." You know, I am going out to the West Coast: I want to say "Hello" to my friends Bonnet and Crossbeak.

What seems to be lacking in much of the literature in . . . ethics in
general is the open admission that we cannot even begin to talk about
the issue of ethics unless we admit that we care (or feel something).
And it is here that the emphasis of many feminists on personal
experience and emotion has much to offer in the way of reformulating
our traditional notion of ethics.[1]

The human capacity to distance ourselves, to separate and to alienate
ourselves, that important precondition for all forms of exploitation, is
paradigmatically expressed in our attitudes toward nature and animals.
Insofar as we can distance ourselves from animals, we can justify
treating them as mere means to our human ends. Insofar as we can
distance ourselves from nature, resist being regarded as part of a larger
order in which humans and the nonhuman realm are reciprocally
interrelated, we can get excited about changing nature all around—
dominating it, harnessing it. . . . This model of humans' relation to the
rest of the natural world . . . [has] produced the worldwide ecological
crisis.[2]

Chapter 4   The "Care" Tradition

Our second lens, which we will refer to herein as the *"care" tradition*, offers
an alternative conceptualization of moral agency and ethical deliberation,
different in many ways from that embodied in the tradition of rationalism's
moral point of view. It suggests, in Diana Meyers words, that "Moral reflection is . . . messier . . . [but] is also more vital and more fascinating than many
philosophers have heretofore thought."[3] The significance of human rationality, the tendency to see the world in dualistic terms, the search for universal principles . . . these and many other elements of rationalism are challenged by this alternative approach to moral philosophy. The assumptions
of the dominant tradition are called into question as representing an incomplete, and in many ways impoverished, understanding of the human experience. And the ethical theorizing that grew out of that tradition is seen as de-
scriptively inaccurate (failing to capture the actual experience of moral
agency) and prescriptively dangerous (encouraging the adoption of a fundamentally alienated and objectifying stance toward the "other").

In an effort to supplement mainstream theorizing, then, care tradition theorists posit an understanding of ethical deliberation that, unlike the moral

point of view, emphasizes the moral agent's embeddedness in the situation and the dilemma's own embeddedness in a particular and highly significant context. Although reason remains a valuable faculty, it is not the "be all and end all" of human existence but rather is enriched by the development of intuition and empathy; this approach adopts the nonhierarchical strategy of integrating reason and emotion and, in fact, avoids dualistic thinking in general. Individual autonomy and the Cartesian dissociation between self and other give way to an understanding of the self as fundamentally in relationship; and abstract universals have less meaning than concrete particulars. Claims of objectivity and impartiality are viewed with skepticism, and ethical dilemmas are constructed in terms of responsibilities rather than duties and competing rights claims. Since this perspective on ethics has been most fully developed within feminist theory, we will devote the bulk of this chapter to an exploration of its formulation there, with special attention to its seminal articulation in the work of developmental psychologist Carol Gilligan.[4]

Unlike the development of extensionism from the rationalist tradition, here the tradition and its environmental ethics "offspring" have evolved largely contemporaneously, and, again, we will find that the environmental ethics theorizing explored in this chapter closely parallels the thinking of the care tradition itself. The alternative approach to environmental ethics has come to be called *ecological feminism* or, simply, *ecofeminism*. The name aptly highlights both the origin and the content of this body of thought: it is situated at the intersection between feminist and environmental concerns, and it explores the central issues that are common to both bodies of theory. More specifically, it sees the oppression of women and the oppression of nature as two sides of the same coin, and it casts both of these issues, in light of its critique of rationalism, as outgrowths of the objectification of the "other" inherent in Cartesian dualism.

Quoting the core and often-repeated phrase of Ynestra King, one of the first to articulate the premises underlying ecofeminism, Jim Cheney summarizes the central insight of all ecofeminist thought, "The claim that the domination of women and the domination of nature are 'intimately connected and mutually reinforcing' is the common thread that runs through ecofeminist writings."[5] Karen Warren, one of the premiere authors in this new field, has offered perhaps its most useful working definition:

> As I use the term, *eco-feminism* is a position based on the following four claims: (i) there are important connections between the oppression of women and the oppression of nature; (ii) understanding the nature of these connections is nec-

essary to any adequate understanding of the oppression of women and the oppression of nature; (iii) feminist theory and practice must include an ecological perspective; and (iv) solutions to ecological problems must include a feminist perspective.[6]

As with feminist thought in general, ecofeminism is wide ranging in its concerns, and thus our discussion here of "ecofeminist care" will only begin to suggest the diversity of thought this field represents.[7]

We will be primarily concerned here with two general topics: the critique of rationalism, which allies ecofeminism with feminist moral philosophy in general and which undergirds the analysis of the dual subordination of women and nature, and the development of the "care ethic" in both inter-human and environmental ethics.

From Feminist Moral Philosophy to Ecofeminism: The Joint Critique of Rationalism

Feminist moral philosophy is a twofold project: it critiques masculine bias in rationalist ethical theory, and it develops its own alternative approach to ethical theorizing, one based on a "reconception of what it to be human and what it is for humans to engage in ethical decision making."[8] The critique of the dominant tradition is rooted in increasing awareness of the discrepancy between the model of moral agency it offers and the experience of women (and, less explicitly, that of other minority groups) in conceptualizing and resolving moral dilemmas. As Eve Browning Cole notes, "many women [have] expressed a longstanding feeling that traditional ethical theories had always strangely seemed to alienate them, to lack something, and to reflect none of the ways in which they themselves think or feel they ought to think."[9] Feminist theory's critique, then, exposes and traces to their conceptual roots those aspects of the tradition of rationalism that have exalted and reified the male perspective while denigrating or ignoring the experience of women. And, second, its development of alternative ethical theory explores this historically dismissed female perspective as the source of important insights into the experience of moral agency. Warren writes, "Sometimes this involves articulation of values (e.g., values of care, appropriate trust, kinship, friendship) often lost or underplayed in mainstream ethics";[10] and, similarly, it may also involve legitimizing predominantly female ways of understanding the self, of reasoning, of dealing with conflict, and of relating to others.

Most fundamentally, feminist moral philosophy shares the underlying impetus of feminist thought in general: "All feminists agree," writes Warren, "that the oppression of women (i.e., the unequal and unjust status of women) exists, is wrong, and ought to be changed."[11] Thus feminist moral philosophy is oriented toward real-world rather than abstract moral questions, and it is self-consciously liberatory in nature; it is critical of the domination of women (and, again less explicitly, of any form of social domination be it racism, classism, or sexism), and it seeks to expose those assumptions of the dominant worldview that implicitly justify oppressive institutions and practices. "Feminist philosophy," in Cole's words, "calls us to the things of this world, and this call is a call for change." It has a problem-oriented, political nature that contrasts sharply with the theoretical, ahistorical stance of the tradition of rationalism.

> Feminist philosophers diverge from the traditional or stereotypical image of the philosopher in . . . address[ing] themselves to particular historical situations, avoiding the flight into abstraction wherever possible. . . . Their philosophical thinking is oriented toward a specific goal: the liberation of human beings from all forms of oppression, foremost among which stands the oppression of women. . . . Feminist philosophers have set themselves the task of dispelling [the] oblivion [of traditional philosophy to gender injustice], by exploring its conceptual roots and transforming their discipline so that it can no longer lack sensitivity to real social injustice, and so that it will no longer be powerless to effect real social change.[12]

Although feminist philosophy reflects the diversity of feminist thought in general, there are at least five key elements or tendencies that seem to characterize much of this literature and that readily demonstrate the difference between the "care" approach and that embodied in the dominant tradition's moral point of view; given the extent to which extensionism shares the defining characteristics of rationalism, this critique is directed toward both mainstream and environmental ethics theorizing.

(1) Feminist moral philosophy challenges the hierarchical, dualistic thinking of Cartesian rationalism.
It is the hierarchical dualisms of Cartesian rationalism that feminist moral philosophers hold responsible for the dominant tradition's explicit male bias and its implicit justification of subordination of the (female) other. As we saw in Chapter 2, Cartesian rationalism is based on a dualistic and hierarchical worldview; the world is fundamentally divided into opposing and mu-

tually exclusive categories—subject/object, male/female, mind/body, reason/emotion, culture/nature, and so on—and the first half of each pair is generally more highly valued than the second. Thus the "male" is set over and against the "female," the "mind" is set over and against the "body," the "public" sphere is set over and against the "private" sphere, and "reason" is set over and against "emotion." The human male is the norm for the thinking subject who surveys and manipulates objects in the world around him and brings about progress within human society, while the human female is equated with the nonrational and the natural, both of which are outside the realm of and beneath male culture.

The initial insight and impetus of ecofeminism was the recognition that women and nonhumans or nature in general occupy basically the same position in this worldview, as inferior "others," and that the treatment each has historically received at the hands of male culture is supported by the tendency of Cartesian hierarchical dualism to render the second, devalued, member of each pair as a mere object. Thus, "rationalism is the key to the connected oppressions of women and nature in the West."[13] Cole explains this central insight: "Ecological feminist philosophers point out that attitudes toward *women* and toward *nature* exhibit an uncanny resemblance; this is not surprising in view of the fact that the central controlling ideologies of our Western tradition define both in terms of their 'otherness,' their 'difference' from the (male) human norm which stands at the center and defines."[14] Two ecofeminist authors, Val Plumwood and Karen Warren, explore this fundamental insight in some depth.

Plumwood: Discontinuity and Instrumental Reason

Plumwood suggests intimate connections between the definition of the human (read male) self as discontinuous from nature (defined as oppositional members of the dualisms) and the tendency to see only instrumental value in the nonhuman/female/nature/body side of the dualisms. Restating the problem with Cartesian hierarchical dualisms in her article "Nature, Self, and Gender: Feminism, Environmental Philosophy, and the Critique of Rationalism," Plumwood writes:

> A dualistically construed dichotomy typically polarizes difference and minimizes shared characteristics, construes difference along lines of superiority/inferiority, and views the inferior side as a means to the higher ends of the superior side (the instrumental thesis). Because its nature is defined oppositionally, the task of the superior side, that in which it realizes itself and expresses

its true nature, is to separate from, dominate, and control the lower side. This has happened both with the human/nature division and with other related dualisms such as masculine/feminine, reason/body, and reason/emotion.

The heart of the critique, then, is what she calls the *discontinuity problem,* or the tendency to define "human" and "nature" oppositionally and then to categorize difference in terms of superiority and inferiority:

> What is taken to be authentically and characteristically human, defining of the human, as well as the ideal for which humans should strive is *not* to be found in what is shared with the natural and animal (e.g. the body, sexuality, reproduction, emotionality, the senses . . .) but in what is thought to separate and distinguish them—especially reason and its offshoots. . . . The upshot is a deeply entrenched view of the genuine or ideal human self as not including features shared with nature, and as defined *against* or in *opposition* to the nonhuman realm, so that the human sphere and that of nature cannot significantly overlap. Nature is sharply divided off from the human, is alien and usually hostile and inferior.[15]

Thus, we see the conceptual linkages between the oppression of women and the oppression of nature; both the female and the nonhuman are set in fundamental opposition to the truly human and are then seen as inferior and therefore subject to the use of the superior human/male.

Warren: The Logic of Domination

Warren has labeled the common pattern underlying the exploitation of both women and nature (and, for that matter, race and class exploitation as well) "the logic of domination." In her article, "The Power and the Promise of Ecological Feminism," Warren claims that the hierarchical dualism of Cartesian rationalism, in conjunction with "the logic of domination" to which it gives rise, constitutes an "oppressive conceptual framework": "a socially constructed . . . set of basic beliefs, values, attitudes, and assumptions which shape and reflect how one views oneself and one's world . . . which explains, justifies, and maintains relationships of domination and subordination." Thus the inferiority of women and nature alike is established and their exploitation at the hands of human males is justified. "It is the logic of domination," she writes, "coupled with value-hierarchical thinking and value dualisms, which 'justifies' subordination."

But what exactly is this "logic of domination"? In its most succinct form, it is stated as follows: "For any X and Y, if X is superior to Y, then X is justified in subordinating Y." Thus, in Warren's argument, the Cartesian-derived justification for the subordination of women and nonhumans runs as follows (reason as the defining trait being one of many possible examples and not the one she uses in this article):

> Men/humans are different from women/nonhumans in that men/humans possess reason whereas women/nonhumans do not (presumed Cartesian fact)
>
> AND
>
> that which possesses reason is superior to that which does not (Cartesian normative claim);
>
> THEREFORE
>
> men/humans are superior to women/nonhumans
>
> AND SINCE
>
> for any X and Y if X is superior to Y then X is justified in subordinating Y,
>
> THEN
>
> men/humans are justified in subordinating women/nonhumans.

What begins as a recognition of difference, then, ends—via the addition of both the Cartesian assumption of the superiority of reason and the statement of the logic of domination—as a justification for exploitative treatment. And, testifying to the broad conceptual significance of Warren's clarification of the logic of domination, this same pattern can be seen to underlie any form of oppression of the "other," be it sexism or speciesism, classism, or racism.[16]

The defining premise of ecofeminism, then, is that these various forms of oppression, both among humans and between humans and nonhumans, grow out of one underlying and fundamental source: the hierarchical dualism-derived tendency toward objectification inherent in the worldview of Cartesian rationalism. Warren summarizes ecofeminism's critical stance toward the two bodies of theory that it unites: "any feminist theory *and* any environmental ethic which fails to take seriously the twin and interconnected dominations of women and nature is at best incomplete and at worst simply inadequate."[17]

And, of course, extensionist environmental ethics has "placed itself uncritically" within the tradition of rationalism.[18] From the ecofeminist perspective, extensionism's reliance on "rationalist-inspired accounts . . . that have been a large part of the problem"[19] manifests itself most clearly in the injunction to continue expanding the moral community to include nonhumans, an approach that fails to address the fundamental problem: it fails to challenge the Cartesian assumptions underlying the objectification and sub-ordination of nonhumans and instead merely imports the same assumptions into its own theorizing. The tendency to define the self and the world in terms of oppositional dichotomies is seen as the source of the problem extensionism is supposed to be addressing, and thus its unmistakable appearance in the environmental ethics theorizing is taken to be highly problematic. Kheel writes,

> the goal of much of the literature in environmental ethics has been the establishment of hierarchies of value for the different parts of nature. It is assumed that hierarchy is necessary to aid us in making moral choices in our interactions with nature. Conflict is taken for granted; it is assumed that one part of nature must always win, while another must always lose. Thus, in a real sense, the field of environmental ethics perpetuates the tradition of dualistic thought.[20]

Plumwood's critique of Taylor's extensionism, for example, emphasizes the self-undermining nature of grounding an argument for biocentrism in the very dualistic, rationalist framework which undergirds anthropocentrism:

> The supremacy accorded an oppositionally construed reason is the key to the anthropocentrism of the Western tradition. The Kantian-rationalist framework, then, is hardly the area in which to search for a solution. Its use, in a way that perpetuates the supremacy of reason and its opposition to contrast areas, in the service of constructing a supposedly biocentric ethic is a matter of astonishment.[21]

Kheel directs similar charges against Singer, Regan, and Callicott. Regan and Singer take for granted that a line will be drawn around an expanded moral community, dividing nonhumans into those are "in" because they meet the criteria for moral consideration and those who are "out," and they take for granted that a central task of environmental ethics involves clarifying principles that govern the necessary trade-offs within this expanded moral community; they "raise the status of [some] animals to a level that warrants our moral concern only to exclude other parts of nature, such as plants and trees." And in a similar fashion, environmental ethicists such as Callicott who argue for the moral considerability of whole biotic communities rather than of individual organisms "are trapped within the dualistic mind set [and cannot] . . . see that moral worth can exist *both* in the individual parts of nature *and* in the whole of which they are a part." Ironically, then, "although many of these writers feel that they are arguing against notions of hierarchy [i.e., those that unquestioningly place the human realm above the natural realm], the vast majority simply remove one set of hierarchies only to establish another"; dualistic hierarchies are maintained in such new forms as sentient animals/nonsentient animals and plants, subjects-of-a-life/non-subjects-of-a-life, and individual organisms/biotic communities.[22]

(2) Feminist moral philosophy replaces Cartesian dualisms with an alternate nonoppositional, nondualistic, and nonsubordinating ontology.
The women's movement and the environmental movement, or more specifically feminist moral philosophy and environmental ethics, are thus united, conceptually and practically, in the attempt to reveal the roots of subordination of the "other" and to displace an inherently oppressive worldview with a set of values that will allow us to replace dualism with recognition of difference and objectification with attentive relation. Caroline Whitbeck addresses this central theme in feminist philosophy in her article "A Different Reality." In feminist ontology, she writes, "reality is understood in terms of the interactions among multiple factors which are often analogous, rather than in terms of dualistic oppositions." Thus the public realm and the private realm, for example, are not mutually exclusive components of society but rather overlapping and interlinked aspects that together contribute to the rich texture of social life. Self and other may be more similar than dissimilar but they nevertheless remain distinct, albeit connected, entities, neither separated by the unbridgeable chasm of dualism nor fused into one indistinguishable whole but rather related; and thus "differentiation [in feminist ontology] does not depend on [the] opposition" that leads to oppression. Similarly, the either-or choice between self and other is seen as a false one, and feminist ontology, unlike Cartesian ontology, can conceive of responding to others in ways that simultaneously nurture the self, the other, and the relationship between the two. The fundamental project, therefore, is not merely the elevation of the devalued half of each pair of Cartesian dualisms but rather the dismantling of hierarchical dualistic thinking itself.[23]

The appeal to human parallels in establishing the criterion of moral considerability in extensionism is similarly seen as insufficiently challenging to dualistic assumptions, so the alternative offered by ecofeminism seeks to call attention to and appreciate the differences between humans and nonhumans. The argument from marginal cases is seen as one of several examples of the pervasiveness of a subtle anthropocentrism in extensionism. Singer and Regan each propose some version of the possession of "interests" as the criterion; the lowest common denominator (as it were) shared by both normal and marginal humans is chosen as the relevant criterion (because any other "higher" trait would exclude marginal humans from moral consideration) and then it is argued that those nonhumans who also demonstrate this trait must, in the name of rational consistency and fairness, also be granted moral considerability. Thus, "Singer and Regan extend the moral community to include animals on the basis of sameness [between them and ourselves]";[24] and this approach is seen as a sign of arrogance and even, once again, as evidence of inconsistent and self-undermining anthropocentrism in a body of theory purportedly attempting to establish nonanthropocentrism. The ways in which nonhumans and the natural world differ from ourselves, the values they may possess uniquely, and the roles they play by virtue of that uniqueness are thus minimized or seen as less important in the realm of ethical deliberation than is their similarity to ourselves, ultimately fostering a devaluation of diversity in itself.

The care tradition in ecofeminism, on the other hand, "involves an ethical shift *from* granting moral consideration to nonhumans *exclusively* on the grounds of some similarity they share with humans . . . *to* . . . [a] structurally pluralistic [stance] . . . that presupposes and maintains difference among humans as well as between humans and at least some nonhumans."[25] Echoing Whitbeck and explaining in this light that the acknowledgment of inherent value in nature, otherwise seen to have only instrumental value, is an insufficient answer to a problem that has been defined too narrowly, Plumwood similarly calls for an understanding of nonhumans on their own terms:

> Challenging these dualisms involves not just a reevaluation of superiority/inferiority and a higher status for the underside of the dualisms (in this case nature) but also a reexamination and reconceptualization of the dualistically construed categories themselves. So in the case of the human/nature dualism it is not just a question of improving the status of nature, moral or otherwise, while everything else remains the same, but of reexamining and reconceptualizing the concept of the human, and also the concept of the contrasting class of nature.[26]

(3) Feminist moral philosophy validates our nonrational faculties and integrates emotion with reason in the process of ethical deliberation.
This third theme, of course, is related to the first in that the presumed illegitimacy of our nonrational faculties is supported by the reason/emotion dualism. Not only is the dominant tradition's "exaltation of the intellect as the core of human life . . . excessive and [in] need [of] correct[ion],"[27] but it is also simply—and dangerously—inaccurate: "We are not (as the Cartesian philosophical tradition might have us suppose) bodiless minds, i.e., 'mental' or thinking beings whose essential nature exists independently from our own or others' physical, emotional, or sexual existence," writes Warren.[28] Posit-

ing a different, more integrated, account of the human being, then, feminist moral philosophy legitimates emotion, intuition, sentiment, and sympathy. It sees emotional response as central to ethical responsiveness, both motivating and helping us to understand the appropriate content of such response. And it refuses to privilege either reason or emotion as components of ethical deliberation but rather "consider[s] them together . . . as aspects of the same process . . . [and thus does] justice to both feeling and thought."[29] Robin Morgan has introduced the term "unified sensibility"[30] to describe the integration of reason and emotion called for by feminist ethical theory.

This issue is discussed in some depth by Kheel in her article "The Liberation of Nature: A Circular Affair." "Most of the literature within the field of environmental ethics," Kheel summarizes, "may be seen as an attempt to establish rationally . . . universal rules of conduct . . . [and most] presumes that reason alone will tell us which beings are of greatest value and, thus, what rules of conduct should govern our interactions with them." Kheel critiques extensionism's appeal to reason alone as a deliberate and well-intentioned but misguided change of direction for the animal liberation movement, away from its early associations with animal lovers who were notoriously driven by sentiment. Driven by the pervasive philosophical notion that "any appeal to emotion is tantamount to having no argument at all," this new direction is less able to capture the full range of ethical responsiveness and less effective in motivating less oppressive behavior because "only those who *feel* their connection to . . . nature to begin with will take an interest in its continuation." Kheel also argues that even the extensionists themselves, unable to fully remove the influence of emotion from their arguments, testify to the fundamental error of the reason/emotion dualism; Singer and Regan's argument from marginal cases, for example, urges us to refrain from mistreating animals by appealing to the otherwise inconsistent protection of insane, comatose, or mentally defective humans, but they do not acknowledge that the appeal is directed to the emotional basis of our unwillingness to avoid the inconsistency by instead ceding the rights of the marginal humans. Thus a "fusion of feeling and thought" is not only a more accurate understanding of our moral response to nonhumans and a more effective motivator of ethical treatment than is the dismissal of emotion in favor of reason, but it has also, albeit without acknowledgment, subtly and inevitably invaded the very extensionist thought that prides itself on maintaining the reason/emotion dualism.[31]

Ecofeminism's development of the care tradition adopts the premise that emotion plays a significant and legitimate role in moral responsiveness and, consequently, argues that environmental ethics "must emerge out of a felt sense of need and personal connection with the issues at hand, not just out of an abstract process of reasoning."[32] As Roger King concludes, "Our emotional attachments to animals, forests, places, landscapes, and ecosystemic dependencies thereby become moral realities to be illuminated by philosophical discourse, not secret yearnings to be hidden and discarded."[33]

(4) Feminist moral philosophy defines the human being in terms of relationships and acknowledges the legitimacy of interpersonal ties in ethical deliberation.
If the challenge to Cartesian rationalism's reification of reason is the integration of thought and feeling, then the challenge to its image of self and other as fundamentally in opposition is the understanding of the self in relational rather than autonomous terms. "What matters," Cole writes of feminist moral philosophy's understanding of the self,

> is that relationships are granted metaphysical priority over isolated individuals, so that the embeddedness of the self in a social world becomes its primary reality . . . ; [thus] the self [is] presented as involved in and importantly constituted by its connectedness to others . . . [and] the relationships in which we . . . stand are [taken to be] of deep significance in defining who we are, how we think, and how we act.[34]

In contrast to the rationalist tradition's casting of the autonomous individual as one whose interests "are defined as essentially independent of, . . . disconnected from, . . . [or] only accidentally [related to] those of other people,"[35] feminist moral philosophy "denies abstract individualism" and posits instead that "relationships are not something extrinsic to who we are, not an 'add on' feature of human nature [but rather] play an essential role in shaping what it is to be human."[36] And, Warren continues, this understanding of the self alters the nature of moral deliberation in that "*how* a moral agent is in relationship to another becomes of central significance, not simply *that* a moral agent is a moral agent or is bound by rights, duty, virtue, or utility to act in a certain way."[37] A major focus of feminist moral philosophy, then, involves clarifying exactly what it means to understand one's self as a "self-in-relation" and exploring the implications of such an understanding for ethical deliberation and ethical theory.

Given this emphasis on emotion and relationship, it should not be surprising that feminist moral philosophy also diverges sharply from the dominant tradition's requirement of impartiality in ethical deliberation. Midgley, for example, states the typical reaction of feminist ethicists to the impartiality of the moral point of view: "I do not think that any case has been made out for supposing that people are *capable* of being emotionally impartial . . . nor for denying that closeness imposes special duties."[38] The critique is therefore twofold: not only does the moral point of view contradict actual experience—in that "there is no value-neutral knowledge accessible to some impartial, detached observer"[39]—but its centrality also renders rationalist tradition theorizing insensitive to the variety of ethical obligations that arise from personal relationships and attachments. In the place of impartial, impersonal deliberation, then, feminist moral philosophy posits an "interpersonal" perspective that acknowledges the ethical relevance of special ties without succumbing to the dangers of deliberation driven by self-interest (against which rationalist tradition impartiality serves, in part, as a defense).

Ecofeminism similarly acknowledges that "relationships of humans to the nonhuman environment are, in part, constitutive of what it is to be a human."[40] It "involves a shift from a conception of ethics as primarily a matter of rights, rules, or principles predetermined and applied in specific cases to entities viewed as competitors in the contest of moral standing, to a conception of ethics as growing out of what Jim Cheney calls 'defining relationships,' [with nonhumans or other elements of the natural world] i.e., relationships conceived in some sense as defining who one is."[41] The relationship between the human moral agent and the nonhuman in question thus helps to shape not only the agent's understanding of herself and of the other but also the nature of her response in the given situation. Ecofeminism "makes a central place for values of care, love, friendship, trust, and appropriate reciprocity—values that presuppose that our relations to others are central to our understanding of who we are. It thereby gives voice to the sensitivity that . . . one is doing something in relationship with an 'other,' an 'other' whom one can come to care about and treat respectfully."[42] Thus "the value we place on particular places or animals should not . . . be relegated to a secondary or 'sentimental' status as exceptions to the universal and impartial claims of morality"[43] but rather should be acknowledged as a legitimate ground for moral response and justification.

(5) Feminist moral philosophy rejects the abstract, generalized representation of the "other" characteristic of rationalist ethical deliberation and instead enjoins attention to context and to the particular details that render each individual moral subject and each moral dilemma unique.

Seyla Benhabib applies the term "generalized vs. concrete others" to this point of divergence between the two traditions. Describing the understanding of the "other" sought by the moral point of view, Benhabib writes that

the standpoint of the generalized other requires us to view each and every individual as a rational being entitled to the same rights and duties we would want
to ascribe to ourselves. . . . We abstract from the individuality and concrete
identity of the other. We assume that . . . what constitutes his or her moral dignity is not what differentiates us from each other, but rather what we, as speaking and acting rational agents, have in common. . . . Each is entitled to expect
and to assume from us what we can expect and assume from him or her.

The mode of deliberation recommended by feminist moral philosophers, on the
other hand, depends on personalized not abstract knowledge of the other; thus

the standpoint of the concrete other . . . requires us to view each and every rational being as an individual with a concrete history, identity, and affective-
emotional constitution. In assuming this standpoint . . . we seek to comprehend
the needs of the other, his or her motivations, what s/he searches for, and what
s/he desires. . . . Each is entitled to expect and to assume from the other forms
of behavior through which the other feels recognized and confirmed as a concrete, individual being with specific needs, talents and capacities, . . . confirm[ing] not only [our] *humanity* but also [our] human *individuality*.[44]

The alternative approach to knowledge of the other developed with feminist
ethical theory, according to Margaret Urban Walker, posits "acute and unim-
peded perception of particular human beings . . . without [which we] really
cannot know *how it is* with others toward whom [we] will act, or what the
meaning and consequence of any acts will be . . . as the condition of adequate
moral response."[45]

Ecofeminism "makes no attempt to provide an 'objective' point of view"[46]
from which environmental ethics deliberation is to occur but rather advocates
decision-making from the very midst of the situation, with attention to the
particular details and the concrete forces at work. "What counts as appropriate conduct toward both human and nonhuman environments is largely a
matter of context," writes Warren.[47] Related to the absence of an objective
perspective is the injunction in ecofeminism's "care" theorizing that, where
possible, we personally experience the situations over which we are making
moral judgments. "When we are physically removed from the direct impact
of our moral decisions—i.e., when we cannot see, smell, or hear their results—we deprive ourselves of important sensory stimuli which may be important in guiding us in our ethical choices," Kheel writes; and she suggests
that "If we, ourselves, do not want to witness, let alone participate in the
slaughter of the animals we eat [for example], we ought, perhaps, to question
the morality of indirectly paying someone else to do this on our behalf."[48]

Feminist ethical theory thus begins with "a rethinking of the 'moral point of view' in light of the social and historical context of human nature."[49] Human experience is taken as the foundation of such rethinking, and the focus is specifically on including in this process previously devalued and ignored aspects of being human. Walker summarizes this alternative approach to ethical theorizing.

> [It is assumed that the] adequacy of moral understanding decreases as its form approaches generality through abstraction. A view consistent with this will not be one of individuals standing singly before the impersonal dicta of Morality, but one of human beings connected in various ways and at various depths responding *to each other* by engaging together in a search for shareable interpretations of their responsibilities, and/or bearable resolutions to their moral binds. These interpretations and resolutions will be constrained not only by how well they protect goods we can share, but also by how well they preserve the very human connections that make the shared process necessary and possible.[50]

Similarly, in uncovering the explicitly rationalist underpinnings of extensionism, ecofeminist authors have focused critical attention on potentially problematic aspects of emerging theory and have highlighted the need for a still deeper critique of anthropocentrism, as Cole summarizes.

> In conclusion, then, feminist ecological philosophy calls for a move beyond the alienated unemotional self, the abstract context devoid of natural human response, the tendency to project human political concepts onto the realm of nature, thereby denying its uniqueness and integrity, and avoiding what Karen Warren has called "logics of domination." To say all this is not to say that there will suddenly emerge simple answers to the enormously complex ecological problems which face us. But feminist philosophers urge that we keep in mind the effects of these beliefs on women and other "others" which mainstream Western philosophy has either created, supported, or condoned. If it has worked to marginalize women, to mystify their lives and concerns, to ratify oppressive social orders, then it will probably *not* work as the basis for an environmental philosophy that can bring us nearer to a harmonious relationship with nonhuman nature.[51]

"Care" and Moral Development Theory: Kohlberg versus Gilligan

The identification of "care" as an alternative moral orientation and thus the impetus for the development of this tradition originated in the field of developmental psychology, specifically in the work of Carol Gilligan. As her work was in many ways a reaction against the dominant model posed by

Lawrence Kohlberg, we will begin this discussion by briefly examining his work, specifically noting its firm grounding in rationalism and its reification of the assumptions underlying the moral point of view.

Having spent over twenty years studying the ethical deliberation practices of a set of male students, Kohlberg proposed what has become the premiere model in moral development theory (study of the process we move through as we mature and become competent adult moral decision makers). Set forth in *The Philosophy of Moral Development: Moral Stages and the Idea of Justice* and *The Psychology of Moral Development: The Nature and Validity of Moral Stages,* the model grew out of analysis of the responses offered by these students to a set of hypothetical moral dilemmas, such as the story of Heinz:

> In Europe a woman was near death from a special kind of cancer. There was one drug that doctors thought might save her. It was a form of radium that a druggist in the same town had recently discovered. The drug was expensive to make, but the druggist was charging ten times what the drug cost to make. He paid $200 for the radium and charged $2,000 for a small dose of the drug. The sick woman's husband, Heinz, went to everyone he knew to borrow the money, but he could only get together about $1,000, which is half of what it cost. He told the druggist that his wife was dying, and asked him to sell it cheaper or let him pay later. But the druggist said, "No, I discovered the drug and I'm going to make money from it." So Heinz got desperate and began to think about breaking into the man's store to steal the drug for his wife. Should Heinz steal the drug?

Discussing such dilemmas with the young men periodically as they grew from childhood to young adulthood, Kohlberg developed a model of moral reasoning structured as a sequence of six stages through which individuals grow as we mature; during each of these successive phases we hold a different understanding of what constitutes right and wrong and a different rationale for obeying the rules of morality:

1. *Obedience / Punishment.* In this early stage we obey the rules of morality out of fear of punishment for transgressions. The right course of action is determined by the physical consequences that we will have to bear as a result of disobedience.
2. *Expediency.* Here obedience is driven by the desire for reward. The right course of action is the one that most satisfies our needs.
3. *Conformity.* During this phase we obey (or conform to) rules in an attempt to gain the approval or avoid the disapproval of others. The right course of action is that which pleases others.

4. *Law and Order.* In this phase we obey rules as a function of acknowledging legitimate authority. Fulfilling our duty determines the right thing to do.

5. *Social Contract.* Obedience in this stage is an effort to live up to standards of the common good. The right course of action is in accordance with public values.

6. *Universal Principles.* Personal integrity and conscience drive our adherence to universal principles in this ideal culminating phase. The right course of action is determined through reason by subsuming the particular situation under abstract principles.

Although we as individuals move from one stage to the next at our own pace and although some of us do not seem to ever progress to the highest stages, the sequence itself is invariant (meaning that the higher stages cannot first be reached without moving sequentially through the lower stages). Note that the underlying assumption of this model is that morality is a rule-driven exercise, that it is grounded in the increasingly rational application of principles. The characteristics associated with the early stages, or with moral immaturity—emotional motivation, partiality, obeying moral authorities—are those that adoption of the MPV encourages us to eliminate from our ethical deliberation, whereas the latter stages bear more resemblance to those elements that we are to strive to achieve if our decision is to have a claim to ethical legitimacy: logical reasoning, impartiality, and universality of principles. The ideal presented in its culminating stage is in many ways the same as that embodied in the MPV: individuals use their highly developed skills of abstract reasoning to apply, from a dispassionate vantage point, universal yet inherently self-legislated principles to determine the right course of action in a given ethical dilemma.

Gilligan posed the first serious challenge to this rationalist model of moral development. Kohlberg himself had noticed that women generally tended to "plateau" at levels 3 and 4 of his six-stage sequence, their moral reasoning even in adulthood evincing concern for the feelings and approval of others more than reliance on the concepts of duty and universal principle. Unlike Kohlberg, however, who tended to accept such results as empirical confirmation of the dominant tradition's long-held assumptions regarding the inferiority of women's moral and rational faculties, Gilligan wondered if the evidence of women's moral immaturity was more a function of Kohlberg's incomplete understanding of the process of moral development. In order to test her hypothesis that perhaps Kohlberg's male-derived schema did not capture the whole story of moral development, Gilligan began conducting research similar to Kohlberg's in its focus on processes of moral reasoning but unique in its selection of primarily female rather than male subjects and in its reliance on actual as well as hypothetical dilemmas (subjects discussed their decisions on abortion as well as the story of Heinz).

During the course of several interview-based studies, and particularly in her discussions with women, Gilligan found explicit support for her hypothesis of a "different voice," a way of engaging in ethical deliberation that was distinct from that emphasized in Kohlberg's work and in the literature of the tradition of rationalism. Gilligan summarizes the distinctness of the two "voices" as heard by Kohlberg and herself:

> The moral imperative that emerges repeatedly in the women's interviews is an injunction to care, a responsibility to discern and alleviate the "real and recognizable trouble" of this world. For the men Kohlberg studied, the moral imperative appeared rather as an injunction to respect the rights of others and thus to protect from interference the right to life and self-fulfillment.[52]

The women's reasoning focused on the particular relationships that they saw as constituting the essence of the dilemmas. They spoke of the abortion decision not in terms of competing claims between their own rights and those of the fetus but rather as a challenge to their desire to respond to the needs of the various parties involved. They defined the ethical problem not in terms of determining their duty by establishing a hierarchy among competing principles but rather in terms of avoiding harming either themselves or others. Gilligan's subjects couched the decision in the context of its consequences for the network of people involved, and their ultimate goal was to find a solution to their dilemma that minimized harm to both self and others and that reflected a balance of the conflicting responsibilities to care for everyone involved.

This alternative moral orientation, which Gilligan has labeled *care*, has two interlinked components.

1. *A conceptualization of the self as interdependent rather than autonomous.* "This ethic . . . evolves around a central insight, that self and other are interdependent,"[53] Gilligan writes, and she reports from her interview data that

> In all of the women's descriptions, identity is defined in a context of relationship and judged by a standard of responsibility and care. Similarly, morality is seen by these women as arising from the experience of connection and conceived as a problem of inclusion rather than one of balancing claims. The underlying assumption that morality stems from attachment is explicitly stated.[54]

The key image in the women's descriptions of themselves as moral agents dealing with the abortion dilemma is "a network of connection, a web of relationships that is sustained by a process of communication."[55] It is important to add, however, that Gilligan did not hear in these self-descriptions the tendency—reflected in the Freudian (male) understanding of the self—to see "an oceanic feeling of fusion"[56] or a merger of self and other as the alternative to separateness; connectedness and attachment occur between *interdependent* selves, not *autonomous* selves or *merged* selves, and thus the women perceived and defined themselves with respect to a third alternative (self-in-relation) that had not been recognized previously by male-derived theory.

2. A mode of ethical deliberation grounded in responsibility to both self and others rather than rights. This conceptualization of the self as fundamentally in relationship fosters and supports a morality of responsibility; just as "a morality of rights and abstract reason begins with a moral agent who is separate from others, and who independently elects moral principles to obey," so too "a morality of responsibility and care begins with a self who is enmeshed in a network of relations to others, and whose moral deliberation aims to maintain these relations."[57] This second aspect of care itself has at least three components:

1. responding to need (such that a young girl considering the Heinz dilemma sees the problem "in the failure of the druggist to respond to the wife");
2. avoiding the infliction of harm (such that a woman considering abortion, frustrated because there seem to be no alternatives that do not cause pain, says, "If there could be a happy medium, it would be fine, but there isn't. . . . It is a choice of either hurting myself of hurting other people around me");
3. resolving conflicts in such a way as to maintain relationships (such that a young girl considering the Heinz dilemma suggests further communication between the husband and the druggist and fears that the husband will be jailed for theft and thus not available to take care of his wife in her illness).[58]

It is important to note that the term "responsibility" takes on a different meaning here than in the detached mode of rights-based deliberation associated with the autonomous self of the dominant tradition: it refers not to fulfilling previously accepted obligations (as in one student's definition, "Responsibility

means making a commitment and sticking to it") but rather to "acting respon-
sively in relationships" (as in another's definition, "Responsibility is when you
are aware of others and . . . of their feelings . . . [and are] seeing what they need
and . . . what you need . . . and taking the initiative").[59]

Gilligan also derived from her data an alternative sequence of moral development. Whereas Kohlberg's stages tie the process of moral maturation
to changing understandings of notions such as duty, rights, equality, and
reciprocity and to an increasingly autonomous self-concept, Gilligan's se-
quential process is based on a maturing understanding of relationships and
of responsibility; the trajectory moves from the pole of self-interest to that
of self-abnegation and then ultimately achieves balance between self and
other in the context of maintaining relationship and avoiding harm. The
five-stage model she posits is as follows:

1. *Individual Survival.* The individual's own needs are the only issue at stake,
   and the right course of action is the one that best (pragmatically defined)
   meets those needs and that ensures self-preservation.
2. *Transition from Selfishness to Responsibility.* The self-interest of the previous stage begins to be criticized as selfishness, and an awareness emerges
   that what one wants to do may not be the right thing to do given one's responsibility to consider the others who are involved; taking care of the
   self is still primary but is understood in the context of responsibility to
   others.
3. *Goodness as Self-Sacrifice.* The needs and interests of others are primary
   at this stage as the individual seeks acceptance and approval through conformity to societal norms and through consensus; the right thing to do is
   to respond to others, even at a cost to oneself.
4. *Transition from Goodness to Truth.* The individual begins to question the
   appropriateness of constant self-sacrifice and considers the possibility of
   balancing responsibility for others with responsibility for the self; although it will probably be condemned as selfish in the eyes of society, including one's own needs in the realm of concern is understood to be more
   honest and more fair.
5. *The Morality of Nonviolence.* Fully cognizant of the hurtful nature of self-
   sacrifice (hurtful to the self, that is) and committed to avoiding hurting
   others as well, individuals elevate an injunction against causing harm to
   the status of an overarching principle that validates the moral equality of
   self and other; care, or not hurting, becomes the universal response to
   moral conflict and its scope includes both self and other.

Using the familiar abortion debate as her illustrative example, Gilligan succinctly and clearly summarizes how the "different voices" of the justice and care moral orientations "each frame the problem in different terms and [use different] moral language [to] point to different concerns," and we quote her here at length in concluding this discussion of her important contribution to moral philosophy:

> The language of the public abortion debate, for example, reveals a justice perspective. Whether the abortion dilemma is cast as a conflict of rights or in terms of respect for human life, the claims of the fetus and of the pregnant woman are balanced or placed in opposition. The morality of abortion decisions thus construed hinges on the scholastic or metaphysical question as to whether the fetus is a life or a person, and whether its claims take precedence over those of the pregnant woman. Framed as a problem of care, the dilemma posed by abortion shifts. The connection between the fetus and the pregnant woman becomes the focus of attention and the question becomes whether it is responsible or irresponsible, caring or careless, to extend or to end this connection. In this construction, the abortion dilemma arises because there is no way not to act, and no way of acting that does not alter the connection between self and others. To ask what actions constitute care or are more caring directs attention to the parameters of connection and the costs of detachment, which become subjects of moral concern.[60]

Although Gilligan originally discovered the "voice" of "care" primarily through the study of women's ethical deliberation processes, follow-up studies (both her own and others') suggest that both justice and care orientations are accessible by men and women alike. Far from being limited to either the female mind or the stereotypically feminine realms of concern, the care orientation is a comprehensive complement to the rationalist perspective, broadening the expression of our ethical concerns and enriching our identity as moral agents; its elucidation has given us a richer and more comprehensive understanding of the human experience in helping us "to recognize for both sexes the central importance in adult life of the connection between self and other, the universality of the need for compassion and care."[61]

Development of an "Ethic of Care" in Moral Philosophy

Several feminist philosophers have undertaken the task of refining Gilligan's sketch of "care" as a moral orientation into an actual "ethic of care." Noting that "in philosophy, there emerged proposals for an 'ethic of care,' conceived as an alternative to the [dominant] tradition's 'ethic of justice,'" Cole suggests the basic elements of this new ethic:

> In an ethic of care, the *relational self* which we have seen as an important ele-
> ment in much feminist thinking today is the basis for moral reasoning. The cen-
> tral directive of an ethic of care is that I should always act in such ways as to
> promote the well-being of both the others to whom I am in relation and the self
> which is relationally constituted. . . . Rules are less significant than a caring and
> attentive conscientious *presence* within one's moral situation, a sensitivity to the
> needs and desires of others, and a basic dispositional willingness to do what I
> can to create situations in which those needs can be met.[62]

Thus, we clearly see how Gilligan's discussion of the "care" orientation to
morality both accords with the five general tendencies of feminist and
ecofeminist moral philosophy examined earlier in this chapter and provides
the glue (as it were) that melds these insights into a coherent and comprehensive ethical system. We now explore two of the primary focal points of
this effort to develop an "ethic of care," particularly noting its distinctness
from rationalist tradition ethical theory, and then turn to an examination of
three examples of the manifestation of the ethic of care in ecofeminist theorizing.

*1. Articulating an ethic grounded in the notion of responsibility rather than
rights.* The first aspect of this project of refining an "ethic of care" that we
will examine involves clarifying what Whitbeck calls "the responsibilities
view" inherent in the ethic. While "the rights view" of ethics "model[s] all
moral requirements on the rights [of individuals] . . . [and views people as]
social and moral atoms, armed with rights and reason, and actually or potentially in competition and conflict with one another," the responsibilities
view of ethics underlying the ethic of care "takes the moral responsibilities
arising out of a relationship as the fundamental moral notion" and focuses
not on accommodating and smoothing the interactions of conflicting rights-
holding individuals but rather on nurturing self, other, and relationship
through appropriate responsiveness to need. While the rights view takes
note of relationships, if at all, only in terms of contracts which give rise to
inflexible obligations or duties, the responsibilities view posits that "relationships between people place moral responsibilities on both parties, . . .
[that] these responsibilities change over time with changes in the parties and
[in] their relationship, . . . [and that] each party in a relationship is responsible for ensuring some aspect of the other's welfare or, at least, for achieving some ends that contribute to the other's welfare or achievement."[63]

Rita Manning, in her article "Just Caring," writes that responsibility in the
sense of responding in such a way as to promote the other's welfare requires

(as Cole suggested above by the term "presence") "a willingness to give the lucid attention to the needs of others which filling these needs appropriately requires," and, although it "can involve a measure of self-sacrifice," it also requires paying attention to and not continually subordinating one's own needs. Manning also suggests, although many feminist ethicists would probably claim that this needs further analysis, that such responsiveness requires neither an emotional attachment to nor an actual relationship with the other; thus it may be that responsibility as the central ethical concept of an ethic of care need not arise from the relationship between self and other, and Manning writes provocatively on this point that "I may meet the one cared for as stranger, though the caring for will change that." And, finally, the actual content of one's response to another in need will vary contextually: with one's ability to help, with the perceived appropriateness of the need in question, with one's concomitant responsibility to others, and with one's own needs.[64]

2. *Exploring the role of empathy in ethical deliberation.* Building on the "responsibilities view" in her article "Moral Reflection: Beyond Impartial Reason," Meyers posits that an ethic of care must acknowledge the central role of empathy in ethical deliberation, for it is *empathy* that "prepare[s] one to respond to people's needs." This line of thought, too, of course, diverges sharply from the dominant tradition's dismissal of the nonrational faculties as irrelevant to ethical decision-making, and, in contrast to the impartial detachment advocated by the moral point of view, it "presupposes some degree of concern for [the other]." In distinguishing "empathy" from "sympathy," however, Meyers draws not on emotional attachments themselves as the basis of deliberation in the care ethic (the only alternative the tradition of rationalism recognizes to impartiality) but rather on the previously noted integration of reason and emotion. Whereas "sympathy" involves actually "sharing another person's feelings" and thus "accepting them as appropriate," "empathy," in contrast, "places a little more distance between people" and involves "construct[ing] in imagination an experience resembling that of the other person." Thus, successfully empathizing with the experience of another person requires us to draw on both our rational and our nonrational faculties in that

> although one draws on one's stock of emotional experience to empathize with another, empathy is defeated if one simply projects one's own characteristic emotional responses onto the other. To empathize well, it is often necessary to hold one's bounteous emotions in check and to mobilize one's powers of attentive receptivity and analytic discernment. Particularly when the other's

background or circumstances are very different from one's own, empathy may require protracted observation and painstaking imaginative reconstruction of the minutiae of the other's viewpoint.

Meyers also touches on the question of the care ethic's extension beyond the realm of personal relationships. Noting that "we usually reserve our empathetic exertions for people we like and with whom we hope to maintain relationships," she also suggests that "nothing in principle bars one from empathizing with someone for whom one feels no affection" as long as the feeling of concern for the other is present (as is the case, she suggests, with psychiatric and social counseling).

Finally, Meyers argues that appropriate responsiveness, as called for by an ethic of care, is impossible without the specific knowledge of the other that comes only through empathetic understanding; since "people cannot act morally without understanding the impact of their conduct on others . . . [and] cannot make recognizably moral choices [without] tak[ing] people's actual feelings into account," empathy is the key to knowledgeably and sensitively tailoring one's response to the person affected. "An empathy-based approach to moral reflection" is thus a crucial component of an ethic of care in that "making empathy central to moral inquiry implies a responsibility to be alert to others' needs and to fashion responses informed by one's empathetic understanding of those needs."[65]

Ecofeminism and the Care Ethic

Echoing the efforts of feminist moral philosophers to develop the "different voice" of relation and responsibility into a comprehensive ethic of care, several ecofeminist authors are engaged in a parallel exploration of the ethic of care as an alternative approach to environmental ethics. Let us take a brief look at three examples of environmental ethics issues—rock climbing, vegetarianism, and animal research—in which ecofeminist authors explore the care ethic.

Warren's rock-climbing narrative and "care"

Warren offers a narrative recounting her experience of rock climbing as an example of the care approach to environmental ethics. After recalling her first day of climbing, in which she "focused all [her] energy on making it to top, . . . climbed with intense determination, . . . and, . . . exhausted and anxious, . . . clung somewhat desperately to the rock," she contrasts this with her very different experience the following day:

I . . . took a deep cleansing breath. I looked all around me—really looked—and listened. I heard a cacophony of voices—birds, trickles of water on the rock before me. . . . I closed my eyes and began to feel the rock with my hands. . . . At that moment I was bathed in serenity. I began to talk to the rock in an almost inaudible, child-like way, as if the rock were my friend. I felt an overwhelming sense of gratitude for what it offered me—a chance to know myself and the rock differently, to appreciate unforeseen miracles like the tiny flowers growing in the even tinier cracks in the rock's surface, and to come to know a sense of *being in relationship* with the natural environment. It felt as if the rock and I were silent conversational partners in a longstanding friendship. I realized then that I had come to care about this cliff which was so different from me, so unmovable and invincible, independent and seemingly indifferent to my presence. I wanted to be with the rock as I climbed. Gone was the determination to conquer the rock, to forcefully impose my will on it; I wanted simply to work respectfully with the rock as I climbed. . . . I felt myself *caring* for this rock and feeling thankful that climbing provided the opportunity for me to know it and myself in this new way.

Thus the human–environment encounter can take either of two forms, both of which Warren experiences while climbing. The climber takes either the arrogant, imposing, nonattentive approach of a conqueror or the approach of a caring partner: sensitive, attentive, interactive, appreciative, introspective, respectful of difference. Note also that the morally significant element in this experience is neither the climber nor the rock itself but rather the relationship between the two; care emerges from the relationship itself, not from "the nature of the *relators* or parties to [that] relationship." And it is through attention to this relationship that the climber comes to experience herself not as a conqueror of the rock but rather as its caring partner.

It is in this transition from experiencing the self as arrogant conqueror to awareness of the self as fundamentally involved in a transformative relationship with a unique other that the ecofeminist ethic of care emerges. "Loving perception of the nonhuman natural world," Warren writes [indebted to Marilyn Frye for the term], is an attempt to understand what it means *for humans* to care about the nonhuman world, a world *acknowledged* as being independent, different, perhaps even indifferent to humans."[66]

Adams and Curtin on moral vegetarianism and "care"

Carol Adams takes this same approach of grounding the theory-building process in narrative accounts, grounding her discussion of the care approach to the question of moral vegetarianism in interviews with vegetarians. Here again, care emerges from and in turn contributes to a deepening awareness of the different types of relationships possible between humans and animals. Adams' interviewees cite their increasing identification with objectified animals as one impetus for becoming vegetarian, and Adams writes of such comments that "identification means that relationships with animals are redefined [such that] they are no longer instruments, means to our ends, but beings who deserve to live and toward whom we act respectfully if not out of friendship." Noting that "becoming a vegetarian after recognizing and identifying with the beingness of animals is a common occurrence," Adams quotes one of her interviewees as an example of both the role of emotions and the awareness of the "other" as an independent subject in ecofeminism's care ethic: "When I thought that this was an animal who lived and walked and met the day, and had water come into his eyes, and could make attachments and had affections and had dislikes, it disgusted me to think of slaughtering that animal and cooking it and eating it." An individual adopting the care perspective on the question of meat-eating, then, might condemn the raising and processing of animals for food, not on the utilitarian grounds of minimizing the pain of sentient beings nor on the grounds of the animals' rights to freedom from imprisonment and pain but rather because "intensive factory farming involves the denial of the beingness of six billion animals yearly" and because she "respond[s] on an emotional level with horror at what each individual pig is subjected to and sympathize[s] with each pig."[67]

The approach of the care ethic to vegetarianism as discussed by Deane Curtin is a "contextual" one: it does not "refer to an absolute moral rule that prohibits meat-eating under all circumstances" but rather admits that "there may be some contexts in which another response is appropriate." Contextual moral vegetarianism does not resolve the conflict between the human interest in eating and an animal's interest in life and freedom by appealing to hierarchical rules (as in Taylor) or to universal, absolute principles (as in Regan) but rather by sensitivity to the context in which the conflict emerges. Thus, since "the point of a contextualist ethic is that one need not treat all interests equally as if one had no relationship to any of the parties," it may be acceptable to kill an animal for food if the alternative is that one's child will starve; similarly, a meat-based diet may very well be justifiable in geographic contexts that render the growing of vegetables impossible. An absolute prohibition against meat-eating is less relevant in either of these cases than are the details of the particular situation. "If there is any context, on the other hand," writes Curtin, "in which moral vegetarianism is completely compelling as an expression of an ecological ethic of care, it is for economically well-off persons in technologically advanced countries"—given that our consumption of meat is by choice and not by necessity, that the factory-farming industry thus inflicts avoidable pain, and that "the injunction to care . . . should be understood to include the injunction to eliminate needless suffering wherever possible."[68]

Slicer on animal research and "care"

In her discussion of animal research from the care perspective Deborah Slicer adopts a similarly contextualist stance. In contrast to Regan's abstract discussion of the issue (largely dismissive of details such as how the experimental animals are treated, what regulations are or should be in place to control this treatment, what gains in knowledge are actually at stake, and so on) and his categorical condemnation of such research, Slicer notes that

> for many thoughtful people the question of *whether* animals should be used in research is more pertinently one of *when* they should be used and *how* they will be treated, just as . . . for many people the question of whether euthanasia is morally permissible is also a question of how and when it is performed, and for some, the question of 'just' war is not so much a question of whether it is justifiable but of how and when. . . . Many people will consider any characterization of these issues that leaves out information about methodology and other contextual features to be decontextualized to the point of being misleading, even irrelevant.

Thus, here it is not the appeal to animal rights that justifies the condemnation of such experimentation as morally wrong, when it is so judged, but rather consideration of factors such as the redundant and trivial nature of much of the research, the unnecessarily painful and invasive nature of many procedures, the avoidably harsh and lonely nature of the animals' captivity, the failure to strengthen and enforce protective measures, and the questionable efficacy of extrapolation from animal data to human health implications. On the other hand, however, an ecofeminist ethic of care will not always, and certainly not simply as a matter of principle, condemn animal experimentation. "Pre-existing emotional or other bonds we might have to members of our own species, community, friends, family, or lovers who may suffer as a result" of any categorical refusal to condone the practice must have their place in the decision-making process; "rather than say that these bonds should count for nothing (as the animal rights literature suggests) or that they count for everything (as the research community suggests)" Slicer's contextualist approach takes them into account, among the many other details of the situation, in the attempt to reach the best decision for all concerned on a case-by-case basis.

And, of course, sensitivity to context also requires personally experiencing that which we are condemning or condoning in our ethical decision-making rather than denying or remaining ignorant of the pain and suffering involved on all sides; thus "those who condone animal research and testing should request a tour of laboratories at the nearest research university [and] should see the equipment—the surgical tables, restraining chairs, 'rape racks,' and 'guillotines'—and experience the smells and sounds" while, presumably, those who condemn it are similarly obliged to experience as fully as possible the costs in human inconvenience and suffering that would (in some, but not nearly all, cases) result from an absolute ban on animal research.[69]

In each of these three accounts of the ecofeminist care ethic, then, we see richer and more complex renderings of the environmental ethics issues than we generally found in the works of the extensionist authors: there is greater attention to the details that uniquely shape each human–nonhuman encounter, there is more willingness to become closely involved in the situation as it unfolds and to experience the full range of intellectual and emotional responses to the dilemma, and there is an acceptance of the "elements of moral tragedy [inherent] in having to make [hard] choices"[70] in situations that simply are not amenable to resolution by clear-cut, absolute, impartial rules. Ecofeminism's care approach to environmental ethics questions captures the "ambivalent antipathy"[71] common to our experience of moral condemnation: our sense that although a certain practice does indeed seem to be wrong, there is at the very least another side to the issue and perhaps even a perspective from which the practice could, in fact, be justified; and, for the most part, it finds satisfaction in our "ceas[ing] to condone practice[s] so cavalierly"[72] without seeking to establish grounds for categorical condemnation.

Potentially Problematic Elements of the Care Ethic

We conclude this discussion with a brief exploration of two aspects of the care ethic which many feminist and ecofeminist authors recognize as problematic.

1. *What exactly it means to be a self "connected" to nonhuman others.* In attempting to clarify the nature of the self as "relational" and "connected," Plumwood contrasts the model of the self found in ecofeminism's care ethic with that found in both the extensionist and the deep ecology approach to environmental ethics. In so doing she follows Gilligan's comparison of the relational self characteristic of the care orientation with the Freud-derived,

male definitions of the self as either atomistic individuals or fused elements of larger wholes. We have seen the retention of the atomistic definition of the self in extensionist theorizing, and we have examined ecofeminism's rejection of this Cartesian-derived model and its alternative description of the self as fundamentally in relationship with, not independent from, others; what remains, therefore, is examination of the distinction between the "connected" self of ecofeminism's care ethic and the "fused" self of deep ecology.

In many ways sharing ecofeminism's critique of extensionism, "deep ecology locates the key problem in human-nature relations in the separation of humans and nature, and it provides a solution for this in terms of the 'identification' of self with nature."[73] Thus its model of the relationship between self and other, at first glance, seems to be that of ecofeminism's care ethic, and indeed the language of connection and relation is used here as well. Plumwood, Cheney, and others have, therefore, attempted to make the distinction between the two models clear, lest ecofeminism come to be mistakenly linked with the different but still "male" perspective represented by the deep ecology alternative to extensionism. In identifying with nature, the self in deep ecology either experiences no boundaries between itself and nature, merging to the extent that there is no independent or unique identity in either the human or the natural realm ("I am part of the rainforest protecting myself")[74] or expands to become an enlarged "Self," which encompasses the natural world and then continues to act out of the familiar strictures of egoism and self-interest ("all that is in my universe . . . is *me*. And I shall defend myself").[75] As Cheney summarizes,

> We see a false dichotomy at work [in deep ecology's alternative to extensionism]. . . . We have either atomistically defined selves who are strangers to one another or one gigantic self. . . . The possibility of defining relationships which, while acknowledging dependencies and bonds of care and responsibility, leave the selves intact simply does not occur.[76]

Returning to Plumwood's discussion, we see that the ecofeminist care ethic, on the other hand, adopts the model of "defining relationships" on the assumption that "we need to recognize not only our human continuity with the natural world but also its distinctness and independence from us and the distinctness of the needs of things in nature from ours." Self-merger is not the only alternative to autonomous, "egoistic accounts of the self as without essential connection to others or to nature," and it is the relational, connected self—the third alternative—that ecofeminism's care ethic attempts to distinguish and defend.[77] Plumwood quotes from Jean Grimshaw:

> Care for others, understanding of them, are only possible if one can adequately distinguish oneself *from* others. If I see myself as "indistinct" from you, or you as not having your own being that is not merged with mine, then I cannot preserve a real sense of your well-being as opposed to mine. Care and understanding require the sort of distance that is needed in order not to see the other as a projection of self, or self as a continuation of the other.[78]

Ecofeminism's care ethic, as we saw particularly highlighted in Warren's rock-climbing narrative, is thus an expression of self "in relationship, not egoistic self [or self] as merged with the other but self as embodied in a network of essential relationships with distinct [nonhuman] others."[79]

2. How an ethic that accounts for moral response to particular human or non-human "others" can nevertheless be broadened to express and inform concern for more generalized or distant "others." A central problem for an ethic of care, discussed by Joan Tronto, is that it seems to abandon those who are "beyond [the] reach . . . of the web of relationships." Given that "we do not care for everyone equally" but rather "care more for those who are emotionally, physically, and even culturally closer to us," it may well be that "there will be some people or concerns about which we do not care"; the question then becomes whether "our lack of care frees us from moral responsibility." If our concern for others, our relationships with them, and our ability to empathize with them actually do take the form of a gradient (increased with proximity, decreased with distance—either geographic or emotional), and it would be hard to argue that in most cases they do not, then an ethic which derives responsibility from concern, relationship, and empathy "could become a defense of caring only for one's own family, friends, group, nation." Noting that "whatever the weaknesses of Kantian universalism, its premise of the equal moral worth and dignity of all humans is attractive because it avoids this problem,"[80] Tronto hints at the growing awareness of this problematic aspect of care tradition theorizing: while avoiding the flip side of the problem, the rationalist tradition's inability to account for partiality in moral response on the basis of special ties, the care tradition, too, has largely restricted its theorizing to only one end of the spectrum and thus, like the rationalist tradition, remains unsatisfactorily incomplete.

A few attempts have been made to deal with this potential shortcoming in the development of the care ethic (although it should be noted that not all feminist ethicists see it as a flaw: for many, it is perfectly understandable and acceptable that we not respond to remote others because we cannot have access to the intimate knowledge of the situation required for appropriate responsiveness and because, in many cases, we have no way to act effectively at a distance). While acknowledging that "Gilligan's empirical work is centered on the domain of personal relations and acquaintances," Blum, for example, claims that "Gilligan means [the] web of [ongoing] relationships to encompass all human beings and not only one's circle of acquaintances."[81] Gilligan herself has suggested that the care perspective, in its "refusal of detachment and depersonalization" and its "insistence on making connections," resists the characterization of being limited in its reach and, in fact, points to "detachment . . . from others [as the] morally problematic [source of] . . . moral blindness or indifference." Specifically, she sees the care perspective as uniquely suited to "see[ing] the person . . . living in poverty as someone's son or father or brother or sister, or mother, or daughter, or friend."[82] Thus an ethic of care can indeed both account for our concern for the stranger living in poverty and encourage response to his need; engaging our empathy and extrapolating from our own experience of relationships, we can come to see the person living in poverty as if he were indeed not a stranger and then respond to him accordingly. Perhaps, then, the superficial boundaries of the care ethic, if there are any, can be expanded through ongoing development of our empathetic abilities so that we are increasingly able to empathize with and appropriately respond to people despite their differences from us or their geographic or emotional remoteness. In any case, at the heart of care theorists' exploration of the boundaries of this ethic is the desire to "defend ourselves against dispositions to keep strangers strange and outsiders outside."[83]

And this (potentially) narrowly delimited nature of particular relationships is equally problematic for ecofeminism's development of the care ethic. Plumwood writes,"normally such essential relation would involve particularity, through connection to and friendship for *particular* places, forests, animals, to which one is particularly strongly related or attached and toward which one has specific and meaningful, not merely abstract, responsibilities of care."[84] But if ecofeminism's care approach is to be a widely applicable environmental ethic, it must surely be able to address human–environment interactions that are not local or particular: we do and should care for and feel responsibilities for "places, forests, animals" we have and will never see, much less develop particular and meaningful relationships with. So the question is whether and how ecofeminism's care ethic can account for and inform these more generalized environmental concerns while retaining its critique of extensionism's abstract universalization; and at least three types of answers are emerging.

(a) Concern for particular others and concern for generalized others are not mutually exclusive.

Plumwood's first attempt to resolve this potentially troublesome issue involves the claim that "the opposition between care for particular others and general moral concern is a false one . . . [which is] associated with a sharp division between (masculine) and private (feminine) realms." Part of the larger critique of dualistic Cartesian thinking, then, is the awareness that although "there *can* be opposition between particularity and generality of concern, as when concern for particular others is accompanied by *exclusion* of others . . . , this does not automatically happen" and the recognition that "emphasis on oppositional cases obscures the frequent cases where they work together—and in which care for particular others is essential to a more generalized morality."[85]

(b) Concern for particular others can lead to more generalized concern.

Thus not only are concern for particular others and a more generalized morality not mutually exclusive, but the former may, in fact, help facilitate the latter. Plumwood writes on this second point:

> Special relationships with, care for, or empathy with particular aspects of nature as experiences rather than with nature as abstraction are essential to provide a depth and type of concern that is not otherwise possible. Care and responsibility for particular animals, trees, and rivers that are known well, loved, and appropriately connected to the self are an important basis for acquiring a wider, more generalized concern.[86]

Slicer makes much the same point about the theory's ability to accommodate both special ties and more extended concern in her discussion of animal research in the context of ecofeminism's care ethic:

> Surely we will not and probably cannot have the same affection for the cat or dog in the laboratory that we have for the animals in our households [any more than we can] have the same feeling for a stranger's lost or abused child that we would have for our own child in a similar situation. . . . While we cannot feel or care the *same* for every human being or animal, the feeling or caring that we do have for our immediate companions should extend some, via imagination and empathy, to our feeling for, our caring about, the plight of more extended others. . . . Someone who has cared about a rock or a tree or a dog or a cat may well care about what happens to, and particularly about the destruction of, other rocks, trees, cats, and dogs. *Such particular relationships can and should enhance one's capacity to empathize, "feel with," and act on behalf of others.*[87]

(c) Contextualized concern need not imply localized concern.
Curtin adds a third important element to this discussion, the distinction between "contextualized" and "localized" concern for others. While agreeing that "caring for resists the claim that morality depends on a criterion of universalizability, and insists that it depends on special, contextual relationships" and although acknowledging that "this might be taken to mean that we should care for the homeless only if our daughter or son happens to be homeless," Curtin nevertheless suggests three ways in which "caring for can remain contextualized while being expanded." This is partly accomplished via the distinction between "the *contextualization* of caring for (the requirement that all caring for has a determinant recipient) [and] the *localization* of caring for, which resists the expansion of caring for to the oppressed who are geographically remote from us."

Significant here, and further developing the point addressed above about concern for particular others leading to more generalized concern, is recognition of a "common context of related interests" that can serve, conceptually, to link particular others with more remote others. Thus, to return to the issues of raising animals for food or keeping them in captivity for experimental purposes, our immediate understanding of the personalities and desires of those animals with whom we share our lives might help us to appreciate the related needs and interests of captive animals and then, in the context of an animal's interest in leading her own life, to expand our concern to those remote and unknown animals as well. Through recognition of this "common context of related interests" we are "enabled to enter into caring . . . relationships that were not available earlier" when concern was merely particular, and in this way the ethic of care is contextualized but not localized.

A second method Curtin suggests for retaining the contextual nature of care without limiting its scope involves recognition of the ties that can come to exist between the self and remote others when those others are affected by one's own choices. "Caring . . . can also be generated by coming to see that one's life (unknowingly) has been a cause of the oppression of others," such that I might experience connection with and concern for those unknown animals whose lives are frustrated by my (mostly indirect) support of the factory farming and animal research industries.

A final element in overcoming the potentially problematic limitation on the realm of care involves, Curtin suggests, drawing a distinction between "caring for" and "caring about": whereas "caring for" depends on immediate relations with particular others, "caring about" "is a generalized form of care that may have specifiable recipients, but [that] occurs in a context where direct relatedness to specific others is missing." Thus we may come to "care about" old-growth forests in the Pacific Northwest upon reading about logging in the newspaper, but we will only "care for" particular trees that we have enjoyed and that are threatened by the logging practices; and our caring for these individual trees may in itself lead us to care about the threatened forest more generally. Ecofeminism's care ethic is thus taken to be comprehensive of both of these distinct aspects of care and therefore to accommodate concern for both particular and general others, albeit in somewhat different forms.[88]

The central challenge underlying these two questions (understanding the self as connected to nonhuman others and expanding the potentially limited realm of an ecofeminist ethic of care) involves making these links between ourselves and nonhuman others more explicit so that we as moral agents have a stronger basis for acting out of care. King succinctly states both the core of the problem and the opportunity for its resolution:

> [We] are increasingly cut off from direct experiential relationships with natural, as opposed to artificial and urban, environments, and thus, although we are unavoidably in relation to the nonhuman world, we do not, many of us, experience that relation to nature as "given" in all its concreteness and complexity. . . . If [as a consequence] people do not care about nature, do not see, feel, or understand it, then an ethics of care is faced with the difficult task of educating the moral imagination to perceive and interpret nature in such a way that nature is consciously a presence in human life, rather than the absence it has become.[89]

Conclusion

We move now to a brief consideration of the ways in which the ecofeminist approach to environmental ethics parallels and extends that of the care tradition to philosophy in general. Just as we concluded our discussion of extensionism in Chapter 2 by noting how it embodied the seven primary characteristics of the tradition of rationalism, so do we now need to reexamine the six characteristic tendencies of the care tradition as they have been imported into ecofeminism's care theorizing. And, again, in so doing we hope to both gain a deeper appreciation of the evolution of thought within this particular philosophical tradition and firmly establish the similarities and differences between these two perspectives on environmental ethics. Beginning with this conclusion but especially when we turn to a retelling of the gray whale rescue in the next chapter, we should be able to appreciate—if not anticipate—how each of the two lenses we have so far examined focuses our attention on different aspects of the human–environment interaction and thus yields different insights and interpretations.

A Nondualistic and Nonhierarchical Worldview

Care tradition theorizing originates in a critique of the hierarchical dualisms of Cartesian rationalism. It holds this worldview responsible for legitimizing the subordination of women and other forms of social injustice, and it sees exposure of the assumptions underlying this worldview as the first step in eliminating unjust practices and institutions. Believing that hierarchical dualistic thinking engenders and justifies objectification of the "other," care tradition theorists supplement their critique with the development of an alternative nondualistic and nonhierarchical ontology, which views difference simply as difference (rather than in terms of mutual exclusion and opposition) and which suggests a more integrated understanding of both self and world. Self and other are thus understood to be fundamentally interdependent or connected to one another.

The original impetus of ecofeminism is the realization that feminist theory itself does not dig deeply enough in uncovering the roots of sexism and thus does not acknowledge the parallel oppression of the natural realm by the human realm. Ecofeminism challenges extensionism on the same grounds that feminist philosophy challenges rationalism: its inherently objectifying and oppressive positioning of those others who are insufficiently similar to the defining model of moral considerability (here, the rational or sentient human; in the more general case, the rational male). As part of the alternative nonhierarchical and nondualistic ontology it proposes, ecofeminism appreciates both the similarities and the differences between humans and nonhumans and sees the recognition of difference, not as the basis for retaining a hierarchy of considerability but rather as the necessary prerequisite for determining the appropriate form of the caring response to the unique nonhuman other. As human beings, we are seen as continuous with rather than in opposition to nonhuman nature, a position less supportive of dualism and instrumental valuation of the other. Further, the human self and nonhuman other are understood to be interdependent rather than atomistically opposed, both contributing to and shaping the larger context of biological life on this planet within which environmental ethics questions arise. Caring for certain nonhuman individuals does not necessitate exclusion of others (as does the extension of rights, for example), and, similarly, value is seen to reside in both individuals and the systems they comprise. And, finally, the project of environmental ethics is not taken to involve an expanded moral community but is rather understood as an attempt to work out the ethical implications of (non-Cartesian) human continuity with the natural world.

The Integration of Reason and Emotion in Ethical Deliberation

The care tradition posits the value of a "unified sensibility": an integration of emotion with reason, which is not only seen as a more accurate rendering of experience but which is also thought to enhance the knowledge gathering and motivation upon which moral judgment and action depend. With its emphasis on integrating reason and emotion rather than relegating either to secondary status, this tradition gives a central place to such nonrational faculties as intuition and empathy; emotional attachments and empathy are seen as both significant motivators and legitimate grounds of ethical response, the model of deliberation is receptive to personal experience rather than merely to the universal demands of logic and consistency, and reason and emotion are seen as complementary rather than mutually exclusive human faculties.

As part of this larger project of reconceiving dualism, ecofeminism's care theorizing defends the role of emotional response to environmental issues and the integration of reason and emotion in ethical deliberation. Arguments for improved treatment of nonhuman animals or for preservation of wilderness areas, for example, need not rely solely on abstract concepts and on appeals to logic and consistency but rather can legitimately acknowledge our emotional attachments and our sympathetic reactions as grounds for behavior change. Sympathy with objectified animals is at least one basis for personal identification with them, and, in turn, such identification often leads us to seek greater knowledge of the situation and then to reject abstract justifications of oppression. Emotional responses provoked by threats to cherished land areas or by the treatment of animals in factory farms and research labs are acknowledged as significant motivators, and receptivity to experiencing our connectedness to nonhuman others on an emotional level is likewise an important part of deliberation. Similarly, experiencing the implications of our decisions as they affect nonhumans is seen as a vehicle for bringing both reason and emotion to bear on decision-making. At the very least, this perspective urges us to move beyond the narrow understanding of ourselves, of nonhumans, and of environmental ethics questions which reason alone gives. Kheel writes of the broader understanding available to reason integrated with emotion:

> Environmental ethics might become more willing to recognize that the fundamental questions about nature and the universe cannot, in the end, be answered rationally. Such an admission may not leave us with the sense of resolution and control that so many of us seem to hunger for, but it may, on the other hand, bring us closer to a feeling of the wonder of the universe and, perhaps, as a consequence, a greater appreciation of all life.[90]

A Focus on Relationship

From the perspective of the care tradition, relationships play a central role in ethics and detachment is morally problematic. The human self is not presumed and encouraged to be autonomous and independent but is rather defined by the relationships it inevitably and enrichingly participates in; ethical deliberation grounded in this understanding of the moral agent as a self-in-relation eschews impartiality as not only impossible but also misguided and instead acknowledges the primary role of the interpersonal perspective. The ethical response begins with awareness of and attention to relationships such that intimate knowledge of the individuals involved, of their relationships to one another and to ourselves, and of their place in the community shapes our understanding of the nature of the ethical response; and relationships both engender responsibility and potentially limit the range of care.

Relationship plays just as crucial a role in ecofeminism's care theorizing as in the care tradition at large, only here the scope of "defining relationships" is taken to include relationships with nonhumans as well; thus, the self is shaped not only by relationships with other people but also, for example, by relationships with particular animals or with particular places. And it is with awareness of and attention to these relationships with nonhuman others that the ethical response begins. For example, it is only when Warren feels herself connected to the rock she is climbing that she is able to move from the conquering to the caring attitude. Ethical deliberation, in fact, in many ways consists of acknowledging the extent and nature of one's many relationships (taking the interpersonal perspective) with both human and nonhuman others and then seeking resolutions that, where possible, maintain these attachments and that respect the unique nature of each relationship. Ethical dilemmas are constructed with reference to the web of relationships affected and not in terms of relative value or competing rights; and "this includes cases where we regularly experience care in return, as in a relationship to a pet, as well as cases where there is no reciprocity, as in the case of working to preserve natural habitats."[91] Cheney makes this point clearly in his discussion of the decision to eat a carrot:

> The question of whether I am justified in eating the carrot is not to be decided by weighing its claims or interests against mine; and this is not because the carrot has no moral standing. The moral issue here is the correct *relationship* to, in this case, one's food. To understand . . . what is required of us, it is necessary to understand the individuals involved (or the nature of the kinds of individuals involved), their relationship to one another, and their place in a complex community or ecosystem.[92]

And, as we noted earlier, it is relationship, not differential moral standing, that underlies limiting the realm of the caring response, particularly where relationship implies the immediacy and knowledge often necessary for effective action; thus, "we may care for the watershed in which we reside rather than (or more than) distant watersheds because we can care more effectively for the one in which we live, not because it has more worth than those other watersheds."[93] The same notion, however, may in many cases oblige us to respond to the needs of distant nonhuman others when a relationship does exist, as perhaps through a chain of oppression linking the agent and the factory-farmed animals she consumes as food, or when our knowledge of and emotional attachments to those nonhumans we are close to (our pets perhaps) give us both sympathy with and ability to respond effectively to similar though remote nonhumans (homeless cats and abused dogs perhaps).

A Contextual Rather than Rule-Oriented Approach to the Resolution of Ethical Dilemmas

From the perspective of the care tradition, ethical deliberation is a context-dependent rather than rule-governed process. The nature of the relationship in question is part of the larger context that in many ways shapes the appropriate understanding of and response to the dilemma at hand. Each situation is to be considered in its uniqueness and in appreciation of the surrounding larger context, and the consistent application of principles or universal rules is neither required nor considered meaningful. By implication, what constitutes an appropriate resolution in one situation may not be the correct course of action in the next; thus moral problems are highly individualized and are not amenable to clear-cut solutions derived from general rules.

Unlike the extensionist authors discussed in Chapter 2, the ecofeminist care ethicists we have discussed in this chapter make no attempt to derive and defend rules or principles to guide the resolution of environmental ethics questions. Instead of reducing such questions to the general characteristics they share and then applying universal rules, these authors assume that the relational and ecological contexts within which each dilemma arises uniquely shapes it and renders it subtly different from every other dilemma; and thus they enjoin individualized, case-by-case consideration of each dilemma. Instead of a general prohibition against meat-eating, then, most ecofeminist care ethicists suggest that while certain conditions may justify the practice (imminent starvation with no other alternative food source, for example), others may not (mere taste preference where nutrition is otherwise adequate, for example). Guidance in ethical decision-making, therefore, comes not from rules but from sensitivity to the details of the particular situation and from attentiveness to the embeddedness of the situation in the broader community. The best one can do is to judge what seems to be appropriate given the types of relationships among the participants in the larger social and ecological contexts, the various needs at stake, and, above all, refusal to dismiss the other as a mere object toward whom the agent can have no responsibilities.

An Understanding of Morality as Balancing Responsibilities

We have seen repeatedly that responsibility and responsiveness rather than the resolution of competing claims is the heart of morality within this tradition; rather than acting out of duty and respecting the rights of others, moral agency is a matter of caring and responding to need and of avoiding harm, and it is failure to respond, not interfering with another's rights, that is morally problematic. Dilemmas often arise because the various needs in question are incompatible and because all resolutions involve harm to some of the parties involved.

Ecofeminist ethical deliberation is not at its core concerned with the adjudication of conflicting rights or interests among humans and nonhumans within an expanded moral community but is rather concerned with balancing the responsibilities that arise in the context of a web of relationships with human and nonhuman others. "Whether or not nonhuman animals [for example] have rights," Curtin states, "we certainly can and do care for them";[94] moral standing is thus not the relevant consideration in our relationships with nonhumans, and in this literature there is very little exploration of nonhuman rights and corollary human duties. What is relevant, though, is the reality of relationships and, subsequently, the importance of maintaining them by avoiding unnecessary harm and by responding to need. Acting as moral agents primarily involves responsiveness in the face of need (rather than noninterference as in the rights view), such that we care for the abandoned cat and try to find him a new home not because his right to life imposes on us a duty to render aid but because we find him in need of assistance that we are able to give.

A Focus on Attentiveness and Appropriate Response to the Resolution of Ethical Dilemmas

And, finally, the model of moral agency embodied in this tradition emphasizes attentiveness and appropriate response to the situation at hand, in light of one's presence in the relationship in question; particularity rather than abstraction is thus the key feature, and the dilemma is often a matter of determining what actually constitutes help versus harm. Understanding morality in terms of fulfilling one's responsibilities shifts the focus of environmental ethics from the extension of moral considerability and the subsequent description of rights and duties to the determination of appropriate responses to nonhumans via attentiveness to the particular details of the context and the particular nature of the relationship. Although the appropriate response to an elderly pet's suffering may be euthanasia, a situation in which a young animal's suffering is remediable may call for intense and expensive medical procedures. And it is only through thoughtful consideration of each situation in all its particular details and subtle nuances that the agent can differentiate between an appropriate caring response and one that, while seemingly respectful, is inappropriate for the individuals involved. As Plumwood summarizes this alternative, and perhaps richer, understanding of human morality enacted in the realm of nonhumans, "With nature, as with the human sphere, the capacity to care, to experience sympathy, understanding, and sensitivity to the situation and fate of particular others, and to take responsibility for others is an index of our moral being."[95]

Sometimes they would come up and set their head up there for a little while and you could really look at them, . . . you [could] see their eyes. . . . It was like they were looking at us. And you could touch them. . . . I would like to have seen them survive after that.

It's very frustrating to see this and to have a feeling that maybe you're just prolonging their misery; maybe there is no way to save them and maybe it wasn't the right thing to do.

Chapter 5 Whale Rescue Story II

We turn now to the retelling of our whale story from the perspective afforded by the care tradition and ecofeminism. Having noted the similarities and differences between the first two of our philosophical traditions, we will gain an even deeper appreciation of their power as conceptual lenses as we compare the versions of the rescue event they respectively produce. This retelling will not only reinterpret many of the elements found in Story I, but will also consider new aspects, omitted as irrelevant from the first perspective, and thus deepen our understanding of why and how the rescue occurred.

If our first lens focused our attention on the rationality (or otherwise) of the decision to attempt the rescue, if it defined the ethical dilemma in terms of the conflict of interests between humans and whales and between individuals and systems, and if it justified the rescue with respect to our duty as moral agents to acknowledge the moral considerability of whales, then how might we expect the version offered by our second lens to differ? As we have seen in the preceding chapter, morality here is not a matter of rationally applying universal principles and acting on our obligations to other members of the moral community but rather involves responding to the needs of those particular beings with whom we are in relationship; emotion and empathy are legitimate complements to reason, and responsibility rather than rights is the central ethical concept. We might expect, then, that this perspective will emphasize the role of relationships with these three particular whales in the instigation and implementation of the rescue effort; we might expect less abstract characterizations and greater attention to the details of the situation and to the broader context, and we might look for evidence of the more problematic aspects of the care ethic: responsibility beyond the realm of immediate relationships and the transition from localized to more generalized concern, for example.

Keeping in mind both the characteristics of our second lens and the version of the story considered earlier, let us return now to our imaginary ethicist and once again pose our four central questions.

Why was the rescue effort the appropriate course of action?

(1) It was grounded in compassionate response to need.
The rescue effort was appropriate because we responded to other beings in desperate need of our assistance. As human beings, we have a "care instinct . . . that responds to hurt and need," and thus, "when the initial decision was made, it was the reaction of 'there's a need, there's a hurt.'" Most of the participants "never really analyzed . . . whether you had a moral obligation to help three whales." Nobody "really consciously made that kind of decision: . . . that we [should do] this because the gray whale is an endangered species . . . [or because] saving those three whales would make a difference in the whale population picture at large." In the words of one participant, "I don't think it was on my mind that it was their right to be saved; it wasn't in my mind that it was an animal rights issue." Rather, "it's more of an instinct, it's more of an emotional reaction: . . . here were three living things trapped and what you want to do is free them"; "most humans just have compassion for something like that." It would have been ethically problematic to have been aware of the whales' situation and in a position to offer assistance and to have refused to help them: "Not to help out some of [our] fellow sentient beings when we had the opportunity . . . [would have shown] that we really have turned into pretty callous human beings." The "care instinct is . . . a human trait," and "you just can't walk away" from other beings in need.

(2) We had a responsibility to help, derived from our having been a cause of oppression.

We also had a responsibility to help the whales that derived from our history of environmental abuse: "the human race has been totally responsible for the elimination of wildlife in its 'care'"; "humans have screwed up the environment . . . badly and when we have a chance to help [animals] out we should." Our having been the cause of such oppression and abuse gives us a special responsibility: "We've almost lost several species. When we have the opportunity to make some rescue attempt, people feel we have this moral responsibility to help them." In this light, the repeated references to the whales as members of an endangered species played a significant role in the instigation and implementation of the rescue effort: "the stranded California gray whales struck a nerve, perhaps because of all the harm mankind has inflicted on whales over the centuries." One observer summed up the source of our responsibility succinctly: "considering the animals' numbers have been reduced because of man's activities in the past, we [should] try to give them a hand in this case."

(3) The relationships between the Inupiat Eskimos and whales were the source of responsibility to help.

The nature of the relationships between the whales and the rescuers also justified the effort, and this is especially true of the "bond . . . that only hunters and hunted understand." If the hunt of the bowhead whale is now more a matter of cultural integrity than physical survival, it is no less the case that the Inupiat people, who "have always lived with these animals," have a "more intense and more intimate" relationship with whales than most; and this "strong feeling of identity with them" dictated their support of and participation in the rescue effort. In the words of one Inupiat spokesperson, "The real essence [of the decision] was [that] our whaling resources provided life and here was one perfect opportunity for us to return the favor to nature." "Inupiat people believe that whales are a gift from [their] Creator," and "this dependence . . . makes [their] actions [in the rescue effort] completely understandable": "no one has more respect for [the] native animals than the local hunters," and "[they] do not like to see animals suffer" so of course "they couldn't just leave them there." Leaving the whales to die was never an option for the Inupiat because, in the words of a spokesperson, "My people are not in that kind of way to treat animals—our renewable resources—that way; if they are doomed to die then we should kill them, so

. . . if there was no alternative for us to help them we were going to shoot them." Once it became apparent that a rescue effort was indeed a possibility, the nature of their subsistence relationship with whales dictated the response of the Inupiat people: "animals that are not killed are respected. . . . If an animal is not needed, it is not killed. And when need be, it is protected, as is the case with the whales."

(4) The relationships between the whales and the people on the ice were the source of responsibility to help.
Other participants in the rescue effort also acted out of a sense of relationship with the whales: an immediate relationship with Bone, Bonnet, and Crossbeak as individuals, a relationship founded in face-to-face encounter and nurtured by daily interaction. Several of the participants commented on the difference actually being with the whales made in their attitude toward the rescue effort and in their judgment regarding its appropriateness: for many of them "initially, no [it did not seem like the right thing to do] but after being there, yes [it did]." One observer expressed the significance of the face-to-face encounter at length:

> It's one thing sitting here . . . being a little detached from the story and kind of saying to yourself "What is going on here?" . . . But to be up in Barrow and then go out onto the ice and . . . see three whales poking their noses through and peering at you with their big eyes . . . and you just touch one . . . it was just very moving: seeing these guys, knowing they were in deep trouble, . . . and that connection with actually touching a whale. Everyone went out and touched one of the whales, and they came away different after that: then you began to root that they would succeed, you didn't want to see these guys drown . . . particularly after you see them and touch them and hear them blow, taking turns at the breathing hole.

"It becomes real when you stand there and you see this living thing coming up gasping for air." Part of the care response is "being there and not being able to walk away from it," and it would be morally problematic "if somebody [could] . . . watch the hole freeze over and let them die when there's a way of keeping them alive."

And "as the thing wore on . . . [people] would kind of become attached or feel empathy for them . . . and really want to get [them] out"; "people kind of started identifying with them after they were around them for a few days," "start[ed] to feel a kind of bond with these creatures." One participant commented on the motivation such a long-term face-to-face encounter with creatures in obvious need provided: "It kept us going [through the long days and nights on the ice that] there we were right up close to these [whales] and you really felt bad for them; they were scraping their noses and they were obviously trapped and they were going to die."

The relationship that developed over time between the whales and the rescue workers and observers was a personal, intimate one. For many people being physically close to the whales, touching them and speaking to them reassuringly, was the "neatest portion of the whole thing"; several noted that the "best day of all" was when they first touched the whales, and others "tried to spend as much time as [they] could out there," even sleeping on the ice in an attempt to "just [be] alone with them." Most of the people who spent much time on the ice with the whales agreed that "they wanted to be touched" and that "they were happy to have someone there who cared about them," suggesting that the relationship was mutual. One woman at the site composed music for the whales and played it to them, and many people spoke with the whales: "They would come up and sort of just sit in the hole and their eyes would just look at you and [people] would say 'Hold on, we're going to get you out, hold on.'" In this same spirit, one participant remembered seeking out the whales after the icebreakers arrived, thinking that "if they're leaving it would be nice to have a chance to say good-bye."

Rescue workers called the whales by their names; their having names "just made it more personal for some people." "In fact, once the animals were given their nicknames, the option of shooting the whales was replaced by the possibility that the defenders patrolling the breathing holes might shoot anything that threatened them, including the polar bears lurking nearby"—a response that many justified by contrasting the depth of the relationship they had with the whales and the lack of a similar relationship with the threatening polar bears. Bone's death, of course, was "devastating" to the rescue workers, and some of them could not bring themselves to admit that he had indeed died but rather held onto the hope that he "held his breath and made it out"; his death, or "disappearance" as many preferred to say, gave the rescue effort a renewed sense of urgency as biologists and rescue workers realized that the remaining two whales could themselves hold out only so long: "after the baby died, you could feel it, that we have got to move, we have got to do something now."

And, not surprisingly, the relationships formed during the rescue effort are ongoing for many people who claim that the whales "will always kind of be a part of" them. "[They] do [think about the whales still] because [they] did get attached to them; [after all] it was everyday, [it] went on for weeks."

In the words of one participant, "I think it would be hard for them not to [still] be with me; I guess I always feel like they are out there," and, similarly, in the words of another, "They are there in your mind, sure."

(5) The relationships between the whales and the general public were the source of responsibility to help.
The desire to help, to respond, was widespread. In their letters, for example, children wrote that "I would like to help but I am in the fourth grade" and "I wish I could help but I can't because I live too far away." Many members of the public felt committed to the rescue effort as the appropriate course of action; although they did not have as immediate a relationship with the whales as did the people working on the ice with them day after day, people across the country and around the world were able to experience the type of relationship that leads to responsiveness through the media. Magazine, newspaper, and especially television coverage mediated this relationship, in effect bringing people far from Barrow into close, personal contact with the three whales: "It was just good fortune that the creatures stranded themselves just fourteen miles from the only television station for hundreds of miles. The image of the three whales, cut off from the rest of their species, trapped by the encroaching ice and gasping for breath, awakened deep feelings of compassion. . . ." "Within a matter of hours [after the original footage of the whales was broadcast] the three stranded whales had the sympathy of the world," in large part because "you could sit in your living room and participate in the thing." The media "brought [the whales] right up close where you could look them right in the eye" just as the workers on the ice could, so it is not surprising that the response to the whales extended beyond the local area: "the whales seem[ed] closer to home [because the public had] seen compelling photos and footage of the familiar gray beasts." Observers around the world thus experienced a virtually immediate, albeit mediated, relationship with these three individual whales who needed their help, so much so, in fact, that "when [Bone] showed signs of pneumonia it was as though someone [they] knew had gotten sick." And the response to this relationship was not only empathetic but also active: letters were written to Congressmen urging their support, funds were sent in to defray expenses, and "whenever [they] needed something, if [the rescue workers] told the media, somebody out there in the world listened and things came."

The rescue effort both grew out of and encouraged a sense of connectedness with Bone, Bonnet, and Crossbeak; they were not distant objects of dispassionate ethical deliberation but rather were beings in need, face to face with us, and we acted—correctly—in accordance with our relationship-derived responsibility to respond to their need. Many of the arguments of the detractors derived from a false conceptualization of the issue, and this was especially true of the ecocentric opposition. "Protecting species in their environment and rescuing individual animals are separate, but not conflicting, issues . . . [and] both are examples of . . . 'humane stewardship of life'; . . . people have a responsibility to save species and their habitats, but it's just as important to 'express an empathetic response for animals as individuals in situations where we see that we can help.'" "The million [dollars] doubtless could have been spent more efficiently for saving more whales. But so what? The money was beautifully spent. In a century that has seen much of humankind at its worst, for a few ticks of the clock we witnessed something better, a time in which God's creatures knew a common bond." In this light, it is perhaps the memorable images of people and whales touching that best captured the spirit of the rescue effort: "[they were] reaching out to touch something else that's living on this planet; . . . for a million dollars the world had the chance to get a little connection with another species for a while, and that's OK."

What was the significance of their being *whales* rather than another type of animal?

(1) We easily form emotional attachments to whales.
The whales' plight was "the focus of . . . the emotions of millions of people of all ages around the globe," in large part because "whales are easy to get attached to." "There is something [about them] almost everybody can like . . . [so that] even if you never saw a whale you would somehow feel connected to it . . . [and] that puts them in a special category." "People generally do feel a more intimate connection with whales than they do with other wildlife." A letter sent to Barrow expressed the popular sentiment, "I love any kind of whales," and a commentator echoed it: "The drama of the battle to save the grey whales off Barrow underscored the emotional attraction that many people feel for the giant mammals of the sea."

The analogies to Jessica McClure, then, were "apt only because of the feelings involved," not because the whales were deemed to share with human beings the traits underlying moral considerability; if anything, the special emotional attachment many of us feel toward whales results not simply from their humanness but from their unique combination of similarities to and differences from ourselves. Part of their appeal is that "there's a mystery about them: they breathe air but live in the ocean, and they're huge but they do seem to communicate"; "their size has a lot to do with it [as well] . . . and there's something otherworldy about them—you don't see them that much—and they're so graceful in the water."

And, finally, the emotional attachment was fostered by the whales' own interest in their human rescuers, by their interactive behavior: "it wasn't an accident . . . that [they let] people touch them; they really seemed to seek out people . . . [and] if their coming up to the people and letting them touch them hadn't occurred, they probably wouldn't have been fought for as hard."

(2) We can readily empathize with them, and therefore we were able to put our-selves in their situation and to identify with their desperation.
For many people, whales are particularly easy to empathize with, perhaps because we are similar in some ways and therefore share some of the same types of experiences. "[In part] because of their intelligence, . . . they do offer a ready vehicle for that personification and identification, where we can really go beyond sympathy to empathy." "It's harder to turn your back on a whale in a breathing hole than on a fish dying right beneath it" be-cause we "sense the intelligence of [the whales] and [we're] afraid they may have emotions—maybe they're afraid." We could "look in their eyes [and] see that they wanted to be free . . . [and] that they knew we were their only hope." For many of the people on the ice with the whales, touch-ing them served as a vehicle for this experience of empathy; in the words of one observer,

> [People] would come back and talk about touching the whales and it truly af-
> fected them. So I went out there and there was this whale sticking [his head]
> up and looking out so I patted him on the nose and it was . . . different: [it told
> me that] here is someone, it's clearly not a rat, it's someone, it's an animal who
> knows what's going on . . . and is aware of its difficult plight. And you just want
> to tell the guy, "They're working hard, don't worry, stay calm, keep alive for a
> while, you'll get out."

And because they were whales—as opposed to, say, rats—and were easy to empathize with, we were able to imagine ourselves in their situation: "[their] predicament . . . [was] easily understandable in human terms." When Bone played with the bubbles made by the de-icers and when he "[came] up to rest himself against the side [of the ice]," we could all em-pathize with him: we know what it is to be young and playful and what it is

to be "tired . . . and weak." "These are young whales [that] have made a mistake," "creatures on a necessary journey" who were probably "mad because
they have to stay ther [sic] so long"; we can relate with these experiences.
Similarly, "people . . . [who] don't like their freedom taken away from them
. . . saw three animals trapped" and easily identified with their fear; "there is
a sort of empathy with something being trapped, struggling to get out."
"There's an identification with an air-breathing creature being caught between the ice; . . . the idea of dying from lack of air, trapped beneath the ice,
creates a pretty strong internal reaction in anyone, any air-breathing creature with emotion."

What was the basis for opposition to the rescue effort?

*(1) A prolonged and potentially futile effort was believed to be actually harmful
to the whales.*
The dilemma was not whether to respond the whales' need but rather what
type of response constituted help rather than harm; thus, some people dis-
agreed that the rescue effort was the right course of action because "the extraordinary measures taken to free the whales may actually prolong the animals' suffering." In the words of one detractor,

> My reaction was "Why doesn't somebody kill these creatures and put them out
> of their misery?" . . . you should put them down like you put a race horse down
> [because] there's no hope here. I wish they could get away, but there's no way
> they're going to get away; even if you get them out of sight they're going to die
> in the ice somewhere. Let's just execute them and save them the grief. . . . If it
> were me, in a similar dead-end situation, I'd much rather have somebody put a
> bullet in the back of my brain when I wasn't looking than to go through this
> agony, slowly approaching death.

Even some of the participants in the rescue, in fact, had similar qualms and
"asked [themselves] would [the whales] have been better off if [rescuers]
had just let them die right away rather than stringing them on": "it's very
frustrating to see this and to have a feeling that maybe you're just prolonging their misery, maybe there is no way to save them and maybe it wasn't the
right thing to do."

(2) The context of natural processes in Alaska suggested noninterference.
For many detractors, the larger context of natural processes in the Arctic
suggested noninterference in a situation that they defined in ecological

terms as "nature's way of feeding other animals." Even some of the supporters of the rescue, in fact, acknowledged that sensitivity to the broader
ecological context and acceptance of life and death as part of the natural
world could render the rescue effort problematic; one admitted, "I have my
own personal feelings about this while recognizing at the same time that
this is a superhuman effort for something that happens all the time" and
another stated, "I'm certainly going to feel upset and very hurt if they can't
make it out; but I'd just have to put it in a wider biological context." In the
context of the Alaskan wilderness, "nature's cruel ways [are] part of life,"
and the ecological perspective acknowledges that gray whales naturally
and probably not infrequently become trapped by quickly shifting pack ice
and either drown or are killed and eaten by polar bears. To the extent that
such deaths are the product of natural selection, the winnowing from the
population of the less-fit genes of animals that are "too stupid" to leave
their feeding grounds before winter sets in, the ecological, contextual response demands noninterference rather than response to need. Ultimately,
"the whales are going to die one way or another," and thus "making a big
deal out of nature's way of feeding other animals" reflects ignorance of or
outright disregard for the broader ecological context within which the
event occurred.

*(3) Frustration with failure of responsiveness in the face of human need was the
source of some of the opposition.*
And, finally, some of the opposition to the rescue effort was grounded in
frustration with our failure to respond similarly in the face of human
needs, particularly given the presumption that closer relationships should
exist between members of our own species than between ourselves and
whales. "Many people [were] struck by the dichotomy between the enor-
mous emotion shown over the whales and the callousness with which we
regard suffering in our own society." Several editorial cartoons expressed
this frustration: one depicted homeless people trying to get help by wearing whale costumes, another showed hostage Terry Anderson (who on the
day the whales were freed spent his forty-first birthday as a hostage in
Lebanon, his third in captivity, while his daughter was broadcast singing
"Happy Birthday") musing that a whale costume might attract the attention of the world to his situation, and a third depicted people dying of starvation in Sudan hearing on the radio that "The world stood transfixed today by the heart-rending plight of the gray whales." These cartoons
captured not a conflict of interest between humans and the whales—"the

whales aren't competing with the homeless women and desperate men of the world . . . [and] the funds supplied by the Guard and Veco and Arco would not have gone to hungry children"—but rather frustration with the too-frequent failure of response highlighted so clearly by the contrast with the whale rescue. "If there was anything to lament in the colossal effort to save the whales, it was not that we care too much for another species, but that too often we care too little for our own." Rather than being seen as in conflict with human needs, then, the whales' needs and our response to them highlighted the failure of response in so many other situations and thus led more than one observer to comment that "Looking around at the Soviets, the Eskimos, the journalists gathered around the whales, I thought the important question was not 'Why are we doing this for these whales?' but rather 'Why aren't we doing this for ourselves?'" And for many detractors, this contrast was even more problematic since our relationship-derived responsibility to respond to need should have been more compelling in the case of human need: as one observer asked, "Do we not have more in common with these wretched human beings—most of whom are suffering and dying through no fault of their own—than we do with our friends the whales?

What are the implications of this event for environmental issues in general?

(1) It suggests both the possibility and the difficulty of building a more generalized concern from such localized concern.
The whale rescue demands that we consider whether concern for these three whales in any way transcended the immediate situation and fostered more generalized concern for whales at large or the natural environment as a whole. Several commentators, in fact, considered whether or not "these whales showed us the way toward global consciousness" to be one of the most important aspects of the event and asked, "How can the human outpouring of concern for three whales . . . be translated into real protection for whales in general?" "If you became sensitized to the plight of those whales you might as a result of that . . . be more sensitive to the plight of whales, the plight of the ocean, the plight of the planet in general." How and when this transference or expansion of concern does or does not occur certainly has serious implications for the environmental community, and the whale rescue may be able to shed some light on this important question.

In many ways it seems that the three trapped whales were indeed "messengers for their kind." In the words of one commentator, "Today there are people all over the country talking about whales who two weeks ago didn't know they existed or thought they were big fish"; in this light, the whale rescue has "got to have a heck of an impact . . . [and] be good for the whales in general." And "one of the most important elements [of the event's positive impact] has been the increase in information" on the part of the public; "common sense would tell you that it led to heightened awareness" because, as one participant noted, "I'm sure there were a lot of people that asked questions: are these really endangered species, how many of these are there in the world, . . . questions that they would normally not ask themselves." Further, the whales' entrapment "shed[s additional] light on the plight of whales in the world" as "a good example of the obstacles they do face on top of commercial whaling." As a direct and immediate result of the rescue effort, then, the public "probably think[s] about whales at large" more than they would have otherwise; the whales served as a "focal point, [and] it's good for people to be thinking about whales [and to ask] 'Why does this matter?'" And because "the pictures still sell, . . . [it seems that] it wasn't limited to just the couple of weeks of the story."

And the implications may have been even broader than fostering heightened awareness and a more generalized concern for whales at large: "Every time we are made more aware that we share this planet with other organisms, it brings us into the web of life." The generalized concern that may have grown out of our caring for these whales could encompass our own species as well as the whales or the environment at large; in the words of one commentator:

> Yes, it was worth every cent to save the two great grays—for our own sake as well as the animals'. When we as humans can learn to value animal life, there might be a fragment of hope that we can stop the environmental destruction of our earth home and have some compassion for our fellow humans who are endangered as well under many circumstances.

Although, from the perspective of one observer, the rescue effort would have been "worth it . . . [simply] if everybody . . . [felt] sympathetic to something in the natural world for a while," the potential implications were much more wide-ranging:

> I doubt these whales took a hunter and turned him around and made him . . . an animal rights activist . . . but I think maybe they threw a little stone in the jar for people who otherwise would not have thought about whales that week.

And there are events in your life that really make you turn a corner: somehow all the little ones add up to what you are. So I see it as positive. I see it for your thirteen year olds; what do they say, those are the people you would affect more than anybody, those kids that are just starting to think. There you have some people who maybe had careers come out of this; . . . maybe people are going to be marine biologists now or Greenpeace [workers] or something all because of a couple of whales in the Arctic.

But this expansion of concern is not a simple matter and is by no means certain to occur. In this light, "what seems odd about this general reaction [to individuals in need] is not that it occurs from time to time but that it goes away so quickly. Weeks from now one will have to be reminded of the whales just as one needs to be reminded of Jessica McClure. . . ." While we would all "like to believe its true" that human compassion and caring is transferable and expandable in this way, there also seems to be "a 'lazy side' to human nature that doesn't want to take the trouble to think about broader problems like Japanese whaling or the treatment of animals generally." For example, although "a postrescue direct mail piece [from Greenpeace] got four times the normal response" and although "there were sizeable contributions made that week or two [to Greenpeace]," "[it's] hard to do, to keep that [concern] going . . . [and their own officials estimate that] it may have had a momentary [impact on their fundraising and membership] but not a lasting one." The primary factor limiting this transition from local to general concern may well be some kind of threshold or limited tolerance for ever-present demands on our compassion. Although the potential is certainly there for increased awareness and deepened sensitivity, "there's misery and tragedy all over the planet, and we do have a threshold of how much we're going to care about"; "people will only take so much and then they just don't react anymore . . . [and thus the result is often] a deafening silence [when these broader concerns are raised]."

(2) It attests to the significance of particularity rather than abstraction in human responsiveness.
The whale rescue also offers insight as to the type of issues human caring seems geared to respond to. "Why, in the face of all the ills that befall living things, should three luckless whales cause such a paroxysm of public concern? One answer is that television could, and did, translate an abstract reality into an immediate reality. 'People actually saw the whales sticking their

heads out of a hole in the ice and crying for help.'" Thus there may be "lessons in the Arctic rescue that could teach us how to generate concerted action on behalf of the other victims we know about but who seem to hover beyond the reach of our empathy." If the question is, "Can we learn to respond to the nameless victims of multifaceted problems as generously as we respond to particular victims of specific questions," the answer from the perspective of the care ethic may well be "Not likely and certainly not easily." The environmental community needs to realize, as was aptly demonstrated by this event, that "you really have to make it personal for people to care."

In this case, for example, "you had to have two whales . . . in order to have anything"; in other words, "there's [a] numbers problem" when it comes to human responsiveness and compassion. "Everybody can comprehend the plight of one trapped child or [two or three] trapped whales. The greater the numbers, the less we care. . . . If there were ten whales trapped in the ice, there would be less fuss. If there were one hundred whales, it wouldn't be a story." Part of this is due to the readier vehicle for identification offered by only one or two individuals: "It's just a fact of human nature that we respond to individuals, and that was reflected in the whale rescue: . . . people can identify with three whales rather than perhaps identifying with the plight of whales in general." The difficulty of identifying with abstractions clearly explains the concomitant responsiveness to these three whales and lack of responsiveness to other environmental and human needs that puzzled so many observers:

> Maybe part of the answer is in the particularity of the whales. Millions who remain unmoved by a generalized Save the Whales campaign were genuinely concerned about these specific whales—just as millions who seem indifferent to the Children's Defense Fund's urging on behalf of poverty-stricken and hungry children were willing to do whatever was necessary to rescue Jessica McClure from that Texas well.

The whale rescue offered "a very powerful image of something to care about . . . specifically, not like the plight of whales and isn't the world going to hell in a handbasket . . . but here's three whales." In the words of one participant, "Seeing the whales: they were real, you could touch them, they were there. You know there are millions of them out there in the water somewhere—and . . . I doubt that we . . . spend a lot of time thinking about that . . .—but these were real, they were there, they were trapped, and so you felt compelled to do something." From this perspective, it made sense that funds would not be withheld from the rescue and diverted instead to the Save the Whales cause in general or to world hunger (for example) because "people don't think that way"; abstractions—whether the equally endangered status of the polar bears that would be shot if they threatened the whales, the health of the gray whale population at large, or homelessness and poverty—do not play as significant a role in our ethical responsiveness as do particular individuals, "real, live, up close and personal." "Here it was very direct; it wasn't . . . some faceless sort of entity out there."

While many environmentalists may well find it "hard not to be cynical when [they] see what is well-intentioned emotion, feeling sorry or feeling sympathy for these creatures, by a world that is totally unempathetic to the state of the environment," they could perhaps benefit from following the example set by other "causes" of motivating concern and action by using particular rather than abstract appeals. The nonprofit sector (raising funds for the arts, for example) and world hunger campaigns have realized that "the checks are usually written when [benefactors] have some sort of personal identification with whatever it is they are giving to" and that "when you see [the face of] that one [hungry] child it hurts and that's how people respond," and they shape their campaigns accordingly. Not old-growth forests, then, but one spectacular old tree or the face of a single spotted owl may be the image that opens the hearts and wallets of the environmentally concerned and that "converts" the unaware and uninvolved.

(3) It demonstrates that the potential to be effective, to actually make a difference, largely determines the range of the caring response.
The third important implication of the whale rescue is that we are more likely to be responsive, to not only feel compassion but also take action, when such caring has the potential to be effective: "People get tired of feeling helpless . . . like they can't have an effect on changing something," and so they engage their "caring capacity, . . . they choose to care, when they feel they can make a difference." One participant explained the impetus to the rescue effort in just these terms: "You feel powerless about a lot of environmental issues and this was something where you felt you could actually do something." "The thought that we could go in and do something" certainly played a major role in the involvement of many of the parties: "it was . . . a clean and neat problem that was solvable when so much involving the

environment today isn't"; "it was a kind of focused struggle that one could win." "There were only three of them . . . and [they were] only six miles from rescue. These are bite-sized statistics, numbers we can understand and absorb, and so maybe affect."

The importance of a sense of potential effectiveness also helps to explain the contrast between our responsiveness to the whales and our failure to respond in the face of other needs: "Part of it may be the clear-cut nature, no matter how difficult, of what needed to be done for the whales. . . . No similarly simple action can save Africans from the encroachment of the Sahara, or the Hutu from tribal war, or ghetto youth from drug-induced violence and despair." Another commentator pondered in this light: "Why is it that a nation's sympathy can be mobilized to accomplish this task, while we live with 35,000 children dying daily? Could it be that helping the whales seems possible to us, while providing health and well-being for suffering children seems like something we are doomed to fail at?" "Sometimes the world is too big and too impersonal and, yes, too messed up even to grasp; we are selective in our concerns out of self-defense. We take on what we can, because we can. It is emotional triage, pure and simple." The lesson for other environmental issues? Much of the public's apparent apathy may well result from emotional overload and a feeling of powerlessness in the face of overwhelming problems rather than from the absence of a basic disposition to care; problems need to be presented in such a way that they appear solvable, and people need to be empowered, to be shown how their individual actions can indeed have an effect. When such conditions as these are met, we may be surprised to find, as many were in the case of the whale rescue, how caring people actually are.

(4) It suggests both the motivational power of emotion and the problematic nature of responsiveness grounded in emotion.

The whale rescue demonstrates the power of human emotion as a motivational force, and for many this was the most significant aspect of the story, as suggested by one observer: "If you took this out and looked at it in a vacuum, you [might] say this is pretty stupid, . . . if you look at it as so many people felt something about this then, that feeling can't be ignored; that feeling is important." "It was clear that our species had an emotional stake in helping these creatures survive," and it was equally well demonstrated that "we don't respond dispassionately to the world around us." "The pathetic scene pierced the human heart like a harpoon," and more than one of the participants linked their commitment to helping the whales to an emotional impetus, claiming "I am an animal lover [and thus] I have a need, a desire to help them." "Their plight touch[ed] a universal emotion in all of us," and even the supposedly objective and dispassionate scientists admitted to "hav[ing] a lot of feelings" for the whales and to being unable to gain a perspective of detachment and objectivity regarding the outcome of the rescue. The environmental community need not shy away from appeals that touch our hearts as well as our minds, as our feelings are powerful motivators for action.

However, the event also highlights potentially problematic consequences of responding to need out of emotional attachment. Because of their frequency, it has become "easy to see images of people being killed; [so] we can just let [human suffering] slide [whereas], for a lot of people, if its an animal it will grab [us] more emotionally, more immediately in some ways" simply because of we are less inundated with the suffering of nonhumans. Desensitization, then, also serves to limit care, and, ironically, it may the inevitable result of attending to the hurts and needs of others; we must assume, in this light, that images of animal suffering will become equally "easy to see" and thus that responsiveness and involvement can, over the long term, lead to detachment. Second, it seems that emotional attachments to and relationships with animals such as the whales carry the potential for overriding attention to and respect for the other's uniqueness; for example, justifications of the rescue were occasionally based on images of the whales as pets: "I think we really cared about the whales. They had been following us around like dogs for two weeks; you have to root for them," and "Would someone let their pet cat or dog suffer and die because its rescue cost too much?" And, third, the rescue also made clear the difficulty of integrating and balancing deliberation and responsiveness: in the judgment of many observers and participants, the rescuers jumped into the effort without a well-thought-out plan and their emotion-driven responsiveness to the whales allowed the effort to "snowball" to proportions that, in retrospect, might be seen as extreme (in this light, one observer noted of the snowball effect that it helped her to understand how easily conflict can escalate into warfare). Given the emphasis on responsiveness, the question of when to say "enough" is a particularly difficult one: in the case of the whale rescue, "Sooner or later, if things [didn't] turn out well, somebody [would] have [had] to make a decision as to when to let nature take its course," but, as several participants articulated the difficulty here, "how do you tell people to stop caring," to become uninvolved within the context of an ethic of responsiveness?

(5) It confirms the effectiveness of visual images in generating concern and motivating response.

And because "photographs hit you on a different—[an emotional rather than a rational]—level," they can play a key role in such motivation. The visual images, consequently, were a "very, very powerful part of the story"; "it was the pictures of these whales that gave this story the impact and the momentum that it had, . . . and that's an awfully good argument for the power of photojournalism." A photojournalist on the site explained the ability of photographs to engage our emotions and our responsiveness:

> We run words through complex computers—our heads—. . . [and although] collections of words will paint pictures for us, by the time our mind sees that picture it has been filtered a lot by the intellectual process. Photographs have the ability to go right in your eye and directly to your heart. . . . Photographs hit you on a very emotional level. . . . To see this majestic animal [like that] is traumatic, and it doesn't go through the filter that says whales die all the time. . . . When you look into their eyes, they don't seem like the dumbest three whales in the world; they seem like unfortunate fellow creatures caught in a problem not of their own making. The photographs were absolutely key to the emotional response to this situation.

"The power of a photograph is [indeed] immense," "allow[ing] you to be immediately empathetic." "Photographs touch us emotionally," they "go right by people's reasoning centers and hit them where they live . . . not the rational construct that we all build to present to the world, but where people really live on some kind of inarticulate emotional level," and so of course "daily TV pictures of [the whales] as they gasped for air in a freezing pool aroused the sympathy of millions around the world." "Pictures [of the whales] really made people think and relate to this problem on a much different level." In the words of another photojournalist on the site:

> I can connect with you with a photo in a different sort of intensity than with words. . . . Because of that one frame of mine there—people touching them— they had a chance to apply some empathy . . . because when you look at a picture . . . you imagine yourself there. . . . I look at this and I can imagine being one of these people [touching the whales]. And I've got a real fact and a real situation, real life, real characters, real clothing, real whales. . . . I kind of put myself into this scene instantly, and I think that's what communicates the empathy: the feeling that this really happened, these are real people. . . . There's an awful power in pictures.

The power of such visual images both depends on and fosters the emotional attachment and empathy that motivate action: "If you don't care . . . [or] have some basic empathy . . . a photograph isn't going to change you, . . . [but it] can help sensitize people to a situation . . . and hopefully make things better as a result." Visual images thus have a significant role to play both in heightening public awareness of environmental issues and in providing a ready vehicle for the empathy and compassion on which personal commitment, responsiveness, and behavior change depend.

An unprecedented opportunity for observation, study, and communication . . . perhaps on their part as well as on ours.

A National Guard Skycrane helicopter, one of the high-tech components of the rescue effort, pulls "Ice Basher" duty.

A historic moment of U.S.–U.S.S.R. cooperation. Rescue organizers fly out to coordinate efforts with the crew of a Soviet icebreaker.

Journalists, rescue workers, and observers on the ice number more than a hundred as the effort intensifies.

Part III Philosophical Tradition III

What stands out most when I recall my experience with Kannik, Putu, and Siku are the whales themselves. They were such magnificent creatures! "Whales are unmatched invokers of awe," and "there was definitely this feeling of wonder" every time I saw them or even thought about them. I never did touch them, though. In a way that was hard because part of me wanted to touch them, "to know in every way I could" who they were and "what they were like." But I knew that the least possible human contact would be the best for them; and "humans fondle everything; they weren't pets for Christ's sake! We want to touch everything and we just mess it all up." It just seemed disrespectful of the whales. I even heard some people bragging about having had a picture taken of themselves touching one of the whales and how that would really impress folks back home, as if they were like those cardboard cutouts of famous people you pose next to rather than whales, wild animals. They "were just a pure curiosity for some people," people who only came to Barrow "because that was where it was at." The children though, they had the respectful attitude: the looks of absolute wonder on their faces when they saw the whales . . . and they never stopped asking about whales, they wanted to learn everything they could about them in school.

But I think that's true of our society at large these days. "The worldwide attention that focused on the three gray whales mirrored growing public interest in the fate of whales generally." "Ecology" is *the* issue now, and whales more than any other animals "have come to occupy a place in the Western heart that is special." "The Save the Whales urge is strong these days": "one" is interested in them and concerned for their survival and not embarrassed to act on their behalf.

Whales have become significant: they are part of our world in a way that lungworts, say, aren't. They matter and we are painfully aware of the threat of extinction they are facing. Didn't someone once say that had human beings gone extinct before the passenger pigeon, the birds could not have mourned our passing? But we can and do mourn theirs; and that gives us a special responsibility: we're the only ones who can understand what extinction, or not Being, means and therefore we're the ones who can guard against it. That puts us in the role of stewards, trying to take care of ourselves and other beings in a way that is not so cavalier or careless of existence. Those three whales "were part of a vanishing wilderness and we hate to add to the demise and therefore we try to preserve as much as possible." "Protecting species and rescuing individual animals are part of the humane stewardship of life which is our responsibility." "That's what makes us human, that we have those impulses to rescue and to save and to feel like we're personally caring."

And this whale rescue really was an impulse. "I don't think anybody sat down and analyzed 'Do I have an obligation to do this?' It was more 'Well, here are these three whales and we've got this equipment and let's help.'" "Truly compassionate people don't think before responding in an emergency situation. They simply jump in and provide what help they can for as long as it is needed." Nobody sat down and weighed their rights against human rights and chose to spend resources on the whales rather than, say, on world hunger. "We just do what needs to be done," we don't deliberate over it. We act; we respond to the situation at hand almost as if we're already involved in it from the start rather than having to decide to become involved. "If it's in your space, then something is calling you to act, and it seemed that wherever we were all around the country, these whales were "in our space," "part of our world." "Just because we are far away doesn't mean we don't care."

Looking back and trying to determine the appropriateness of the rescue effort in light of our role as stewards is difficult, however. All too often when we interfere with nature, try to step in and manage it, such action goes hand in hand with not respecting it, not letting it be what it is on its own terms. In the minds of many people, rescuing the whales "was like going out and freeing Bambi," and if that was our attitude toward them maybe the effort was inappropriate. "The whole point of whales is that they're big and wild and free, that in their own dominion they are as powerful as we are in ours. They are not big, fat people who swim well. They're whales. To Bambi-ize or human-ize the noble beasts of the sea shows a lack of respect for what whales are all about." So perhaps truly respecting their Being as whales called for noninterference in this case. But, on the other hand, while respecting the Being of other beings may sometimes require standing aside and letting them take care of themselves in accordance with their own natures, it may also require "reaching out to help them in times of great emergency." "What was the alternative? Watch them die there or let them die their own death out there in the ocean where they are at home and there's always the chance they will survive." I guess it comes down to whether we were protecting them for us or for themselves, making them into a cause so that we could feel good about ourselves or appreciating them on their own terms.

Anyway, what it all comes down to for me is this: "We finally know that human life is tied to the life of plants and animals. We know, better than we did, that we are part of nature, not disembodied from it and not enthroned above it." "We are all beings on this earth"; nonhumans "are not things"— a whale is a being in and of and for itself—and we are not "masters and possessors of nature" but rather shepherds or stewards. We are "connected to each other and to the world and so we responded to the whales on a deep level." "We would have felt a part of us had died" if we didn't at least try to help the whales, and for that reason this rescue effort taught us some valuable lessons about ourselves and our place in the world: "One might forget that people are driven as naturally to preserve as to destroy were it not for the occasional endangered animals who raise their heads from the lower depths and cry for help."

Hast thou ever raised thy mind to the consideration of EXISTENCE, in and by itself, as the mere act of existing? Hast thou ever said to thyself thoughtfully, IT IS! heedless in that moment whether it were a man before thee, or a flower, or a grain of sand? Without reference, in short to this or that particular mode or form of existence? If thou has indeed attained to this, thou wilt have felt the presence of a mystery, . . . a sacred horror, . . . a something ineffably greater than [thine] own individual nature . . . which must have fixed thy spirit in awe and wonder.[1]

Humanity has become blind to [the fact that] . . . human existence is not the master of entities, but rather is in the service of the self-manifesting of entities. Although he decentered the modern subject, which conceives itself as the ground for all meaning, purpose, and value, [Heidegger] nevertheless emphasized the uniqueness of human existence. Yet, he criticized anthropocentrism because he regarded human existence as authentic only when serving a disclosive purpose that transcends it. For him, then, . . . the environmental crisis is a symptom of a still deeper crisis: a humanity made arrogant by its blindness to what it means to be human.[2]

Chapter 6   The Phenomenological Tradition of Martin Heidegger

We move now to the third and final lens of our study, the *phenomenological tradition* as embodied in the work of Martin Heidegger. As the opening quote suggests, the approach undertaken in this body of theory is rather different from that in either the rationalist or the care tradition, and in some ways it is more difficult because of its seeming unfamiliarity; more so than in the previous chapters, certainly, we will have to be patient with new terminology. This unfamiliarity is merely superficial, however, as it results from phenomenology's calling to our attention aspects of our experience that we generally take for granted and thus have little practice in naming or acknowledging. On a deeper level we will find this perspective to be at once familiar, intuitive, and resonant with our own everyday experience; "it is, after all, ourselves [Heidegger] is describing, yet us from the inside."[3]

On one level, Heidegger can be read as less than revolutionary with respect to environmental philosophy in that, as George Cave notes, he "offers radically new interpretations of the Being of inanimate objects and of man, while remaining silent on the problem of the Being of nonhuman animals."[4] Increasingly, however, environmental philosophers are exploring phenomenology in general and Heidegger's work in particular, and primary among these authors is Neil Evernden. Calling attention to the often indirect way Heidegger speaks to environmental philosophy in his text *The Natural Alien*, Evernden represents the spirit of this renewed attention to phenomenology: "Heidegger says relatively little in a direct fashion about other creatures, but by describing us to ourselves he opens the way to an understanding of our consequences to others."[5] Insofar as "the issue [of environmental ethics] is ourselves among others,"[6] then, the field of environmental ethics can indeed be informed by the phenomenological tradition which in many ways focuses on this very question.

A brief comparison with our two previous frameworks may help clarify the nature of this third tradition. Phenomenology does not emphasize ethical deliberation per se but rather takes as its central question the experience of human existence. In our first lens, the rationalist tradition and its offspring, extensionism, the problem at the core of the human–environment interaction was understood to be an inappropriately (narrowly) bounded moral community, and the solution involved expanding the moral community to include (at least some) nonhumans and then applying the standard rationalist approach to ethical deliberation (adopting the moral point of view). In our second lens, the care tradition and its ecofeminist offspring, the central problem was seen as an inappropriately rationalist understanding of the self and of morality, and the answer involved granting legitimacy to nonrational aspects of the human experience, exploring the ethical implications of defining the self as fundamentally in relationship, and acting out of "care" for particular nonhuman others in relational contexts. Each of these traditions was subtly shaped by its underlying assumptions regarding the human being as either autonomous, rational, and principled or relational and responsive; and it is here that phenomenology can be seen as engaging in a parallel project. Heidegger understands the central issue as an inappropriate "mode of being," an inappropriate understanding of what it is to be-in-the-world, an inappropriate conceptualization of the relationship between the human self and others; and the answer involves the realization of a more appropriate mode of being. Although not strictly a body of ethical theory but rather a philosophical exploration of the experience of being human (thus falling into the branch of philosophy known as *ontology,* or the study of existence itself), the phenomenological work of Heidegger nevertheless has ethical implications for human–human and human–nonhuman relationships.

Our exploration of this third tradition will follow the established pattern: in this chapter we will look at Heidegger's philosophical theorizing, focusing on its implications for environmental philosophy, and in Chapter 7 we will return to the whale rescue to see how it might be recast in light of this third lens. Before beginning our discussion of Heidegger, let us turn first to an introductory overview of the broader phenomenological tradition.

The Phenomenological Tradition

The phenomenological tradition originated with Edmund Husserl in Germany in the 1920s. Although largely overlooked in American universities until the recent turn toward postmodernism (with its realization of the extent to which reality is "constructed" rather than given), phenomenology has long been one of the dominant bodies of philosophical thought in Europe.

Phenomenology is the study of phenomena. But by the word "phenomena" we mean not (as it is popularly understood) UFOs and ESP—things for which we have no satisfactory explanations at present—but rather those things that *are* and thus can be experienced, or, in Heidegger's words, "that which shows itself in itself."[7] Turning to Heidegger's interpreters for help with a definition, we find that "phenomenology is . . . a stripping away of concealments and distortions, such as will let us see that which lets itself be seen for what it is."[8] These concealments and distortions include the presuppositions we bring to our encounters with the world around us: for example, that rationality is a better guide to understanding than is intuition, that physical objects are more "real" than emotional states, or that humans are superior to nonhumans. Phenomenology strips away these a priori (formed before the encounter) preconceptions or interpretative constructs and takes the position that "whatever is encountered simply is, without qualification."[9] From this starting point, phenomenology explores various ways "the world is" as rendered through various and unique modes of human experience; the meaning of our experience of ourselves and of the world and how our experiences create the world around us are the primary questions.

Evernden explicitly and comprehensively links the phenomenological perspective to the questions of human–environment interaction. Phenomenology, he writes, "challenge[s] . . . those assumptions about the world" that environmentalists seek to alter and focuses attention on the experiential understanding of ourselves as interrelated with other beings; instead of "defin[ing] relationship to nature out of existence" as the Cartesian worldview does, phenomenology acknowledges that "the constitution of relatedness is an essential part of being human, and that we conceal or repress it at our peril."[10] Edward Relph also asserts the relevance of this tradition to questions of environmental ethics: "phenomenology leads to self-awareness and a heightened sense of responsibility for the environments we live in and the ideas we express about them . . . [and] helps people to appreciate their worlds and lives more fully."[11] Here we will explore four of the central characteristics of this tradition.

(1) Phenomenology challenges the assumptions of Cartesian rationalism.
The original impetus underlying Husserl's development of a new philosophical perspective was his sense that rationalism "decapitates philosophy" by focusing only on the facts and thus excluding as meaningless the most important questions about human life. In its place "he began to formulate a way of thinking about philosophical questions that did not make assumptions about what could be studied and how but which was responsive to all phenomena of human experience."[12] This new way of thinking also discarded Cartesian presuppositions regarding the essential separation between the thinking subject and the world of objects and instead asked of the human subject how he actually experienced his relation with the world around him. Evernden paraphrases Husserl's reaction to Cartesian rationalism and thus highlights the distinctly different assumptions on which phenomenology is based:

> Consciousness is always consciousness *of*. There is no such thing as consciousness as an isolated entity, only consciousness of something—not even Descartes can just think. The object is inevitably built into the act of consciousness, and that which is perceived is not dependent solely on the object out there, but on the actual manner of grasping in consciousness. Thus . . . it is misleading to speak of an isolated self surveying a world, for the person is from the start *in* the world, and consciousness is always *of* the world. . . . By shifting attention from the reality of the world, which Descartes was forced by his assumptions to attempt to prove, to the *meaning* of the world, phenomenology disposes of the distance between the thinker and his object. And given this different point of departure, one is free to ask quite different questions.[13]

And it is precisely this asking of different questions and this challenging of Cartesian assumptions that the environmental movement must, but generally and ironically does not, engage in. The environmental movement, according to Evernden's phenomenological perspective, is torn between the "official" version of the (instrumental) value of the natural world, which has permeated the very science and policy arenas within which environmentalists work, and the personal experience of (inherent) value as attested to by environmentalists themselves. The Cartesian understanding of our relationship with the natural world is as follows:

We know that we are, we exist, and that beings are just other entities that we encounter and learn to master. . . . Of course there are subjects and objects, of course there is man and there are resources—it is obvious. Descartes made it obvious and secured us from involvement in the world. In convincing us that the world is composed of distinct subjects and objects he insulated us from concern with the world and made it next to impossible for us to regard the world as anything but a storehouse of material.

In accordance with this official version, environmentalism ironically takes the form of resourcism, and self-interest becomes the primary motivation for environmentally sensitive behavior. Preservation arguments are flavored with economic, conservation overtones that ground appeals in human interests. Cost-benefit analysis is the premier tool of environmental policy-making understood as wise resource management, and, in this light, "environmental degradation is not necessarily wrong, it is just imprudent"; trees are described and valued as "oxygen-producing devices," the preservation of a swamp is justified "because it serves us by detoxifying wastes," and "even . . . landscape beauty is . . . transformed into a resource and described quantitatively." As Evernden summarizes, "Once adopted [environmentalism as] resourcism transforms all relationships to nature into a simple subject-object or user-used one"; nonhumans and the natural world are understood to have instrumental value only and thus are "preserved" only for future use and only insofar as their utility remains. The saving grace of such an approach, of course, is its effectiveness, at least in the short term; and many environmentalists, quite understandably given the urgency of their concerns, have "succumbed to the temptations of expediency."

And, from the phenomenological perspective, "therein lies the fatal weakness of the . . . movement"; "the act of justifying the [swamp] as a glorified septic tank entails acceptance of the very scale of evaluation which is the environmentalist's most formidable adversary." Environmental phenomenology shows us that environmentalists "have lost precisely what [they] set out to defend—[the natural world's] right to an independent existence and recognition of [its] inherent worth." Phenomenology's key contribution to environmental philosophy, then, lies in its discovery of this ironic adoption by environmentalists of "the strategy and assumptions of their opponents," in its recognition that the official version not only underestimates but, in fact, "completely . . . misunderstand[s] the kind of value experienced by a person in the presence of nature," and in its legitimation of the true goals and values of environmentalism that are submerged and even denied in the context of Cartesian resourcism. It reminds environmentalists and environmental ethicists that the heart of their cause is a "defence of meaning," that "they are protesting not the stripping of natural resources but the stripping of earthly meaning"; and it legitimates this protest in its undisguised, non-Cartesian articulation.[14]

(2) Phenomenology grounds theory in actual experience and especially legitimates subjective experience.
More specifically, though, what is the alternative approach to knowledge that we must adopt in order to begin asking these different questions and then ascertaining meaning in the world around us? Noting that "subjective encounter, the very attitude that is disposed of in [Cartesian] method, is taken to be fundamental in phenomenology," Relph suggests that

> rather than treating the world as somehow independent of us, it requires that we reflect on our own consciousness of things and explore the various manifestations of them in our experiences. . . . In other words, phenomenology insists that I must think about how I experience the world; my dispassionate assessments, frustrations, imaginings and emotional outpourings are all considered to be facts of experience, and one fragment of my experience is not arrogated into some special status. In fact, quite the opposite is required; I have to learn not to see objective measurable facts as somehow more important than my own experiences.

As an example of this alternative approach he considers the phenomenon of time. While from a Cartesian perspective "time is [a] unidirectional, . . . constant frame of reference," from the phenomenological perspective "time is . . . variable and multidirectional for we have memories and moments of prescience; there are days which hang heavily and weeks which fly past"; thus "from this point of view it makes sense to say of somewhere nerve-wracking and hideous, like London's Heathrow Airport, that 'I spent several days there one afternoon.'"[15]

And in paying heed to actual experience, phenomenology is also able to offer an alternative, true-to-experience, version of environmentalism. The phenomenological perspective, according to Evernden, confirms that "the environmentalist's experience does not fit with the official view of the world . . . [as] neutral matter for transformation and exploitation" and then takes this discrepancy as the starting point in offering the movement a way out of the "trap" of resourcism. In contrast to the official version, this model of human–environment interaction does not "trivialize the experience of living and assert the reality of a valueless world . . . [but instead] springs from an understanding of the world that . . . accord[s] with . . . individual experience."

Accordingly, environmental phenomenology pays heed to efforts, in the old style of naturalists and Romantic poets, of concerned individuals to "evoke in [their] listeners a sensation reminiscent of [their] own in the presence of nature," and it does not try to hide the fact that "the incentive to preservation [is] personal and emotional, . . . a reflection of the value experienced by certain persons in the face of nature." It gives a central place to testaments of wonder such as John Muir's well-known encounter with the orchid: "I felt as if I were in the presence of superior beings who loved me and beckoned me to come. I sat down beside them and wept for joy."[16] Phenomenology encourages openness to and appreciation of beings in and of themselves; and this in turn leads to a renewal of "irrepressible wonder at the existence of life" and to concern at the prospect of annihilation. "Relieved of the [Cartesian] context which declares that *this* is important and *that* is not, . . . one is simply aware of what is," writes Evernden, precisely capturing the essence of Muir's encounter with the orchid buds in the wilderness; "wonder is the absence of interpretation" and is thus the almost inevitable result of a phenomenological approach to the world. And, of course, wonder is an important aspect of environmentalism, for to experience wonder is to be open to new ways of understanding the world and to deny the sterile categorization of beings as mere "things." Thus, not only is the phenomenological project grounded in attention to actual experience, but it acknowledges the direct relationship between limited experience of nature and uncritical acceptance of the official interpretation and thus encourages direct experience with the natural world as one of the necessary components of moving beyond the Cartesian model of detached exploitation.

By heeding our actual experience of nonhuman nature and constructing an account of the human being and the world accordingly, phenomenology "removes the Cartesian roadblock" to acknowledging value independent of ourselves and to challenging our instrumental and exploitative relation to nature, precisely the obstacle that environmental philosophy seeks to overcome.[17]

(3) Phenomenology seeks to understand the world from the perspective of the one experiencing it.
It may be helpful here to look at another real-world example of the contrast between Cartesian and phenomenological approaches; in the following case, originally from Erazim Kohak and discussed in Evernden's *The Natural Alien*, take note of how the elements of Husserl's original phenomenological

insights regarding the connected rather than detached nature of human consciousness come to life:

> After a few puffs, the subject [a smoker] looks anxiously for a place to deposit his ashes. There are no ashtrays. The subject casts about, settles on a seashell or a nut dish, and, with a mixture of anxiety and relief, knocks off the ash. He did not "find" an ashtray "in the objective world"; there was none there to be found. Rather he constituted an ashtray in his act.

A Cartesian interpretation of this case notes a thinking subject distantly surveying a collection of objects, one of which he uses as an ashtray, and ignores the experience by which the subject shapes the world around him; "the thing identified . . . seems more central and more real than the act . . . —the act remains effectively invisible for it is not noticed." A phenomenological approach, in contrast, focuses on the intimate connection between the subject and the world around him, on the subject's discernment of significance in that world, and on the act of consciousness whereby the ashtray—as such—came into being; "to be ignorant of the act," Evernden suggests, "is to miss the very context of meaning for the object . . . [for the ashtray], not there at all until it was required, . . . exists where it functions, in [the smoker's] experience." Phenomenology thus attempts to *describe* phenomena in such a way that reflects the understanding of the persons experiencing them; as Evernden summarizes the approach, "Once a phenomenon has been selected for study, one must examine each occurrence of it with a view to grasping its meaning in the experience of the person concerned . . . [and then, cumulatively] the meaning of such experience to mankind in general."[18]

Environmental phenomenology extends this commitment to understanding the world from the perspective of the one experiencing it to include the perspective of nonhumans. It challenges the Cartesian assumption that only human beings can be understood as subjects—beings with intentions, desires, "worlds" of meaning—and that, consequently, nonhumans can only have the status of mere objects. Phenomenology's critique of Cartesian presuppositions thus includes a challenge to the discontinuity it posits between human and nonhuman and an alternative to the "restricted definition of animality" that validates our perception and treatment of them as mere "things" with only instrumental value. From this perspective, animals have individual experiences of and perspectives on the world; they have their own independent existence, and it is only in denying this that we perceive them as objects, existing only for us. And once we acknowledge an animal's subjecthood, we are able to adopt, at least partially and in accordance with the extent and depth of our knowledge, that being's perspective. The classic example is that of the tick: "the world of the tick [like our own] is basically a set of gradients of . . . significant features, [in this case] . . . light, sweat, and heat. . . . [And] to speak of [the tick only in terms of] reflexes and instincts is to obscure the essential point that the tick's world is a world, every bit as valid and adequate as our own." Not only does the tick inhabit a subjective world just as we do, but we are able to put ourselves in the tick's place and thus to perceive his world on his own terms, an aspect of the phenomenological perspective with wide-ranging implications for our interactions with nonhuman animals.[19]

(4) Phenomenology calls us to conscious experience of our everyday lives.
And we too are enjoined to make this "presuppositionless return to the things themselves"[20] in our everyday lives. While "in their broadest sweep phenomenologists give us an overview of the experience of being human,"[21] we will see in our exploration of Heidegger's thought that those who work in this tradition also recognize its potential for encouraging us toward a different—a more aware and more appreciative—understanding of our own individual existence. On this issue of the relevance of phenomenology in enhancing and altering our individual lives, Relph concludes that

> Phenomenology . . . leads us back to the ways in which we relate to others and the meanings we struggle to see in our own existence. . . . Phenomenology is an attempt to see clearly. If we can achieve such vision it is bound to change our attitudes to our own existence, to other people and to the world around us. It will make us more aware of the meaning of our lives, more critical of theories and ideologies which restrict the quality and variety of existence, and, to use a term of Heidegger's more inclined to 'spare', that is, to avoid imposing our wills on things, environments or other people. This attitude of "touching the least" requires that we develop a sense of caring and responsibility for things, places and people that allows them to be themselves with the minimum interference from us.[22]

Evernden concludes his discussion of the contrast between Cartesian and phenomenological perspectives and of the relevance of the phenomenological perspective to environmental ethics as follows: "The sense of separation which Descartes bequeathed to us may well be the most potent adversary of environmental thought [and] our next task, therefore, must be to examine the tactics of [phenomenologists] who have tried to challenge [it]."[23] This challenging of the Cartesian split between self and other is precisely

Heidegger's project; he gives us the grand overview of the human experience and calls us to a wondering appreciation of the world and thus a renewed commitment to our relationships and responsibilities within it. And so we turn now to an examination of his work, focusing on the ways in which it speaks specifically to the question of our relations with nonhuman beings and with the natural world as a whole.

The Philosophy of Martin Heidegger

Martin Heidegger is recognized as one of the "greatest and most creative philosophers of the twentieth century."[24] In the spirit of Husserl's critique of rationalism, his text *Being and Time* (1927) refutes the relevance of the central issues that have preoccupied western philosophers throughout history and suggests a new path for true "thinking": a path guided by a revised understanding of human existence. The goal of his work was not simply to provide a necessary corrective to philosophical thought but, additionally and relatedly, to warn of—and perhaps to ward off—the dangerous consequences of the influence of rationalism on western civilization: the alienation and dehumanization characteristic of the technology-driven, consumption-oriented society of the twentieth century and the unreflective, uncritical acceptance of and adaptation to technological "progress" that reinforce these threats to our lives, our world, and our very being.

The Question of Being

The central question in Heidegger's thought concerns the meaning of *Being*: Being (with a capital "B") or existence itself as opposed to "beings" or things that exist. He thought that philosophy had been on the wrong track ever since Plato and Aristotle; in their efforts to understand the basic structure and properties of the world, philosophers from the very beginning had overlooked the most fundamental fact of all: that the world *was*, that it existed. For Heidegger, the most significant of all questions the Greeks might have asked was "What is the Being which renders possible all being?"[25] or "Why beings at all, rather than just nothing?"[26] Because the earliest philosophers overlooked the essential issue of Being, the very inception of philosophy was inappropriately marked by a focus on beings rather than Being.

It is thus not surprising to Heidegger that "the subsequent history of Western philosophy has been one of the forgetting of Being."[27] Paralleling our own earlier positioning of Descartes as the inheritor and refiner of Platonic rationalism, Heidegger sees the problematic dualism of the Cartesian tradition as the inevitable culmination of the trajectory of thought that legitimated the study of beings while ignoring the more fundamental questions of Being. Evernden summarizes this discussion from *Being and Time:*

> Once our "forgetfulness" [of Being] had taken hold, it was only a matter of time until its consequences were thought through and the world-view implied by it was made explicit. "I think, therefore I am" may seem a significant insight, but it avoids a more important one. Descartes took for granted the verb. He concludes: "I *am*"—but what does it mean to be? "With the '*cogito sum*' Descartes had claimed that he was putting philosophy on a new and firm footing. But what he left undetermined when he began in this 'radical' way, was the kind of Being which belongs to the *res cogitans* [the thinking thing], or—more precisely—the *meaning of the Being of the 'sum'.*"[28]

And along with the rationalist worldview we have inherited has come a tendency in our language that is so deeply ingrained and so pervasive that it is difficult to recognize: our constant use of the verb "to be" without any meaningful consideration of the Being to which we thus refer. Heidegger asks us to stop and think through what the verb "to be" actually means, or, as Evernden puts it, to ponder "What is 'is'?"[29] Rather than focusing, as we generally do, on the concepts of "cat" and "black" in the sentence "The cat is black," we must instead note the more primary and, when we stop to think about it, astonishing fact that the cat "is" in the first place. We are then confronted with the frightening but ultimately powerful and liberating questions that we otherwise struggle to avoid: what it means to be, why it is that we or anything else should exist at all, and how and why it is that we now *are* but someday will *not be.* It is in our very ability to ask these questions, in fact, that our distinctiveness as human beings lies, and it is in struggling with them that we come to realize the possibilities and responsibilities inherent in our being human. Heidegger invites us to experience, consciously and deliberately, "the sense of awe aroused when man becomes aware of the mystery of Being—'that wonder of all wonders, that the beings *are.*'"[30] Openness to this facet of human experience is of crucial importance for Heidegger because, as Evernden summarizes, the "single-minded preoccupation with beings," which pervades not only philosophy and science but also the worldview within which we conduct our everyday lives, "has dramatically altered our understanding of the world and ourselves, and has jeopardized both."[31]

The Human Being as *Dasein,* the Shepherd of Being

Heidegger takes as his starting point for the investigation of Being that particular being who is concerned with Being as such: the human being. Although we are only one among many expressions or embodiments of Being (anything that "is" shares this characterization with us), we are the only beings who are aware of Being and of the possibility of not-Being. Heidegger's word for the human being is *Dasein,* "a being for which, in its Being, that Being is an issue."[32] Being is an issue for us; it is something that we are consciously aware of and something that we can and do ponder. "[These] concrete, inescapable question[s] which everyone has concerning how he is to understand and decide about his own existence"[33] Heidegger calls "*existentiell*" *questions.* Evernden summarizes the importance to Heidegger of our ability to ask these questions about Being and hints at the ethical considerations implicit in our being *Dasein:*

> Heidegger treats this as a virtually diagnostic feature of being human—the awareness of the possibility of non-being. Our own mortality makes that unavoidable, much as we may try to side-step it. But there it is: not only might we not have been, but we were not, and we will not be. We are a temporal phenomenon, a momentary flash of being soon to fade to non-being. But what might the consequences of that flash be, however brief? What might it illuminate? What is revealed through the circumstances of human being? And what is the significance of there existing a creature for whom Being is an issue?[34]

One of Heidegger's own students objected to this definition of the human being as a major inconsistency, accusing Heidegger of "perpetuating the anthropocentrism and dualism so characteristic of the metaphysical . . . traditions which he purported to overcome."[35] Laura Westra, however, represents the position of many (although by no means all) of Heidegger's interpreters in stating that the understanding of human beings as *Dasein* "does not place man in any sense 'above' other things."[36] Heidegger himself refers to the "illusion" by which "man exalts himself to the posture of lord of the earth"[37] and insists that *Dasein* is the servant of Being, not the master of either Being or beings. Heidegger interprets the significance of our uniqueness in terms of the stewardship it suggests rather than translating it into a justification of domination. John Macquarrie makes just this point and, in doing so, emphasizes the consequent relevance of Heidegger's perspective to the concerns of environmental philosophy:

> Man remains unique. . . . But the notion of autonomous man yields to the notion of man as the responsible steward, and perhaps this is by far the more

mature notion, and one that has great relevance as technological man increasingly subjects the world to his control and has its resources more and more at his disposal. Is he the absolute master, or is he rather the one to whom Being has graciously entrusted itself and on whom it has conferred an almost frightening responsibility? Heidegger's answer is clearly given: "Man is not the lord of beings. Man is the shepherd of Being."[38]

Far from undergirding anthropocentrism, then, the acknowledgment of our uniqueness as that being for whom Being is an issue is, in fact, the key element in overcoming our arrogant disregard for the Being of other beings, for when we see ourselves as the unique being *Dasein* we understand our role as shepherds rather than masters and we transcend the Cartesian belief that all beings exist for us.

Heidegger's phenomenological study of the human being as *Dasein* involves describing "the basic structures of human existence as these are disclosed to us in our own existing."[39] Among these basic structures experienced by us as *Dasein* are the following: (1) our being is "in-the-world" and (2) our way of relating to the world is through "care." The remainder of the discussion of Heidegger's philosophy will explore these two central aspects of human experience.

Our Being Is "in-the-World"

As *Dasein* human beings are, first and foremost, *in-the-world*. Once again, perhaps the best way to understand this phrase is to contrast it with the rationalist view of the human being. For Descartes human beings are *res cogitans,* "thinking things," and are thus defined by our separateness from the rest of the world; the moral point of view itself, in fact, positions us, in our most fully human capacity as rational moral agents, above and beyond the objects of our ethical deliberation. Constitutive of Heidegger's *Dasein*, however, is a fundamental relation to the world that denies this image of the self as an independent subject in a world of objects. We noted earlier the phenomenological challenge that "not even Descartes can just think" without there being a world of which such thinking is a part, and here again we see that "man is inconceivable without a world to which he already stands in relation."[40] It is meaningless, in other words, to speak of a person standing alone, separate from the world and from other beings, defined only by his independent will and without essential reference to other beings. "The world is always one that I share with others [in that] the world of *Dasein* is a with-world [and] being-in is always being-with others," writes Heidegger;[41]

"there is no such thing as a man who exists singly and solely on his own."[42] Heidegger's term for the human being, in fact, captures just this sense of our essential relatedness:

> Dasein is the word for human existence *in its ownmost and most proper way of being,* . . . [and] as such the word *Dasein* describes the fundamental comportment or relationship that "humans" have—to the world. . . . This names the fundamental shift, in Heidegger's thinking, away from subjectivity and its objectifying, *to* the always already relatedness that cannot be objectified.[43]

Despite the similarity, this being-with is more fundamental than the particular relationships of feminist moral philosophy:

> According to Heidegger, "being-with" is a basic feature of *Dasein's* being, more basic than relating to particular others. Even when I am not encountering others . . . others are there for me. I have a readiness for dealing with them. . . . Being-with is an aspect of being-in-the-world that makes possible all encountering of particular others.[44]

And just as our being-in-the-world contradicts the Cartesian model of the self as fundamentally separate from the rest of the world, so too does it invalidate the Cartesian presupposition of the rational subject cognitively surveying a world of things: we do not experience ourselves essentially as "knowers" in a world of things to be perceived and quantified but rather as "concernful doers," participants intimately and actively involved in the world.[45] "In everyday terms," writes Heidegger, "we understand ourselves and our existence by way of the activities we pursue and the things we take care of."[46] Thus we are "in" the world in the same sense that we might be "in love": it is not a spatial but an experiential "in," signifying involvement and concern. Heidegger makes the point clearly: "*Dasein* . . . is nothing but . . . concerned absorption in the world."[47]

Heidegger emphasizes the difference between viewing the world as a collection of things to be perceived (known) and participating in the world as the setting for concerned involvement by distinguishing between the *present-to-hand* and the *ready-to-hand*. The former phrase denotes the understanding of the world in terms of "thinghood," which we experience when we draw back from involvement in it, objectify it as a collection of knowable items, and then assign value to it in accordance with its usefulness. The latter term denotes the understanding of objects that "always already" have significance in the context of our concerned participation, the interpretation we experience in our active involvement in the world. Although we do engage in both modes of relating to the world, the mode of understanding things as "present-to-hand" is derivative and secondary, not primary as Descartes thought. George Cave paraphrases Heidegger on this point to the effect that "concernful involvement in the world rather than knowledge of it is the primordial truth of man's way of Being-in-the-world," and he notes of this distinction that "it is only when *Dasein holds back* from any kind of concernful involvement with the world that [he] can encounter the world and entities within it, through perception, as merely present-at-hand, as objects for a knowing subject."[48]

This understanding of *Dasein*'s being-in-the-world in terms of concerned involvement rather than detached knowing accounts for two specific aspects of actual experience that will be important in our discussion.

(1) Our tendency is to perceive spatial distance subjectively (as was noted of time earlier).
While the Cartesian subject, distantly surveying a world of things, places importance on the objective measurement and calculation of distance, *Dasein* experiences *de-severance* (literally, undistancing: think here of how close we sometimes feel to our loved ones during times of celebration or sadness even though we are not with them, or of how distant we feel from them during a prolonged disagreement even though we may be in the same room), determining "remoteness and closeness relative to its concernful absorption."[49] Heidegger writes: "'De-severance' amounts to making the farness vanish—that is, making the remoteness of something disappear, bringing it near. *Dasein* is essentially 'de-severing': it lets any being be encountered nearby as the being which it is."[50] Thus, we experience nearness or distance as a function of the significance of or our involvement with an object or entity and do not necessarily define it objectively.

(2) Our tendency is to "act first and think later" when responding to the world.
Whereas the detached Cartesian subject seeks to know and thus realizes himself, first and foremost, in *thinking, Dasein*—primarily at work in the world, absorbed and involved in *doing*—experiences conscious deliberation only secondarily and derivatively: only when, for whatever reason, his activity stops and his involvement gives way to detached consideration, analysis, or planning (as in the common expression "If we stop to think about it"). "If knowing is to be possible as a way of determining the nature of the present-to-hand," Heidegger writes, "then there must first be a deficiency in our having to do with the world concernfully."[51] Most fundamentally, then, we act: we respond

to, we take care of, we engage with, we participate in. And the situation itself generally reveals what needs to be done, minimizing the actual need for deliberation, both at the point of becoming involved and during the course of involvement. We deliberate only when our concern is "deficient" or when ongoing involvement demands reflection or theoretical consideration, and even then such deliberation "is not the pure detached theoretical reflection described by the [Cartesian] tradition . . . but rather . . . must take place on the background of absorption in the world."[52] One consequence of *Dasein*'s tendency to act without recourse to deliberation is worth noting: because responding to the situation at hand does not depend on justificatory reason-giving or on abstract rationalization in terms of ultimate principles (that is, we do not stop to marshal justifications in defense of our involvement and we do not deliberate whether or not to respond in terms of the values at stake) *Dasein* is more receptive than is the Cartesian ethical deliberator to discovering meaning and significance within the bounds of the situation itself.

And, finally, what exactly is this "world" that we are "in"? It need not be thought of (although it can be) as the physical planet Earth; rather, the world is the region, in an experiential sense more than a geographic sense, of significance. The world is "constituted by significance," and the beings comprising it are those objects and entities that are "connected to one's concrete concerns at any instant."[53] Relph offers a helpful description of this phenomenological understanding of the "world" we are "in":

> We live in a world of buildings, streets, sunshine and rainfall and other people with all their sufferings and joys, and we know intersubjectively the meanings of these things and events. This pre-intellectual world, or life-world, we experience not as a set of objects somehow apart from us and fixed in time and space, but as a set of meaningful and dynamic relations. That is to say, other people, objects, types of scenery, architecture and places matter to us to greater and lesser extents; we are concerned about them and we care for them.[54]

We come close to this concept of a "world" in our use of such phrases as "the business world" or "the academic world" in that we mean to denote the physical environments, the issues of concern, the types of relationships, and the norms of behavior that go along with the context of business or scholarship. These specific "worlds" are "'modes' of the total system of 'equipment' and practices that Heidegger calls *the* world," as are the "public world" we all share and our own individual ("domestic") worlds. Thus, although "my world" consists of those objects, people, animals, events, practices, places, standards, experiences, sensations, explanations, and so on, that have a place of significance in my life, it is important for Heidegger to clarify that "there is no such thing as *my* world if this is taken as some private sphere of experience and meaning, which is self-sufficient and intelligible in itself, and so more fundamental than the shared public world and its local modes."[55] As Heidegger concludes, "The surrounding world is different in a certain way for each of us, and notwithstanding that we move about in a common world."[56]

Our Way of Relating to the World Is through "Care"

Fields of care

In his attempt to "give us the 'essence' of how we are," Evernden writes, Heidegger describes us as "a being for whom Being is an issue and whose way of relating . . . is through 'care.'"[57] To begin our discussion of this second basic structure of human existence, let us recall the notion of de-severance: *Dasein*'s tendency to relate to objects of significance in his world not spatially, in terms of objective distance, but rather experientially, in terms of concern. "Concern," Heidegger writes, "decides as to the nearness and farness . . . [such that] whatever this concern dwells with . . . is what is nearest."[58] Heidegger does not want us to envision here, however, a subject standing in the center of his world mentally bringing objects closer to himself (a Cartesian image); in Dreyfus's paraphrase, "*Dasein* must be thought of as pure concern, not as a physical body located at a certain point in objective space."[59] Heidegger's own words suggest an understanding of *Dasein* as a region or field: "If *Dasein*, in its concern, brings something nearby, this does not signify that it fixes something at a spatial position with a minimal distance from some point of the body. . . . Bringing-near is not oriented towards the I-thing encumbered with a body, but towards concernful being-in-the-world."[60]

Evernden explicitly introduces the term "field" for the understanding of *Dasein* Heidegger articulates here: "If we could conceive of a 'field of care' or 'field of concern,' we might have a means of gaining a partial understanding of Heidegger's description of human being. His term is not 'field,' however; it is 'Dasein' . . . and 'the Being of Dasein itself is to be made visible as *care*.' . . . We know *Dasein* by the evidence of care."[61] And William Barrett, also struggling to adequately express Heidegger's understanding of *Dasein* as concerned involvement in-the-world rather than as mind within a physical body, offers the following interpretation of being-in-the-world-as-care: "My Being is not something that takes place inside my skin; . . . my Being, rather, is spread over a field or region which is the world of its care

and concern. . . . [I] secretly hear [my] own name called whenever [I] hear
any region of Being named with which [I am] vitally concerned."[62] Evernden explores the implications of this Heideggerian understanding of ourselves as fields of concern in-the-world of significant nonhuman beings:

> This may not be altogether unlike the experience of persons who are moved (of-
> ten to their own surprise) to defend the useless, non-human world around
> them. Each "secretly hears his own name called whenever he hears any region
> of Being named with which he is vitally involved." Whether it is the housewife
> who defies the chainsaws to rescue a tree that is beyond her property yet part
> of her abode, or the elderly couple who unreasonably resist expropriation of
> their home, or the young "eco-freak" fighting to preserve some vibrant, stink-
> ing bog, or even the naturalist who fears the extinction of a creature he has
> never seen, the phenomenon is similar: each has heard his own name called,
> and reacts to the spectre of impending non-being.[63]

"Care" versus "technology"

For Heidegger, care is *Dasein*'s way of "revealing" or "disclosing" the world.
As a less-than-accurate but perhaps helpful analogy, consider care for a moment as an attitude and note how our attitudes can affect the way the world
around us appears: if I am in a bad mood, if my attitude is a mean-spirited
or negative one, then I am likely to perceive my husband (and everything
else) as a troublesome annoyance—I "reveal" him (and the world at large)
as annoying and troublesome—whereas when I am in a good mood and have
a more positive attitude I am apt to perceive him as the loving, attentive, generous person he is—I "reveal" him as loving, attentive, and generous. We
can thus translate what Heidegger calls "revealing" or "disclosing" beings in
the world as "interpreting" them, "making sense of" them, or "discovering"
them; and we see that such "revealing" can occur in different ways.

Caring for others involves what Heidegger calls "making them present," or
revealing them as the manifestations of Being they are. As care we are open to
the Being of another being; we disclose it as what it is in and of and for itself.
An important component of care is the attempt to sincerely understand another
being's existence from its own perspective; "revealing as care," we "think" its
being and thus try to "experience . . . how it is that [Being] manifests itself" in
that particular being.[64] Care thus involves paying careful attention to other beings, guided and driven by the enhanced appreciation that comes from intimate
knowledge rather than by the intent to turn knowledge to manipulation. It is
in this "heedful caring for the things that make up our world"[65] that we, in Heidegger's term, "dwell" (or live "in regard to Being") in-the-world as *Dasein*. As

our analogy with attitudes suggested, however, care is a bit more complex than this; and we can clarify what care is by contrasting it with what it is not, with a mode of revealing that is fundamentally "not-caring."

Our revealing of other beings is affected by our society's general understanding of these beings, by the interpretations of *das Mans,* "the They" or "the One." As Heidegger writes, "The dominance of the public way in which things have been interpreted has already been decisive even for the possibilities [of being affected]—that is, for the basic way in which *Dasein* lets the world 'matter' to it."[66] In other words, our understanding of beings and thus our revealing of them is governed, as is most of our behavior, by the social norms embodied in what "they" say or in what "one" does: think here of that indefinite "they" out there who tell us that we must get eight hours of sleep each night, eat from the four food groups, and open doors for ladies, or of how we all know that "one" must pay one's taxes and that "one" does not yell "Fire" in a crowded theater.

Under the sway of the Cartesian worldview, the dominant public interpretation, contemporary *Dasein* tends to reveal beings as resources, interpreting their Being in terms of their usefulness to us, rather than "making them present" as what they are in and of and for themselves; "we experience everything including ourselves as resources to be enhanced, transformed, and ordered simply for the sake of greater and greater efficiency."[67] It is in accordance with this generally accepted understanding of beings that the world comes to "matter" differently and, thus, that *Dasein* can be seen as "not-caring." As we saw earlier with the distinctions between present-to-hand and ready-to-hand and between deliberating and acting, however, this alternative mode of relating to or revealing the world—(mis)understood by Descartes as fundamental to the human subject—is actually only secondary and derivative. Although it is indeed possible for us to relate to the world in accordance with the Cartesian understanding of beings (just as it is possible for us to draw back from the world and seek objective knowledge of things as present-to-hand), our primary Being, our "ownmost and most proper way of being" is that of relating to the world through care.

Heidegger calls this alternative way of revealing the world "technology." It is important to keep in mind here that Heidegger does not use the term "technology" in its usual sense: "For Heidegger . . . technology is not simply the utilization of tools but the understanding of the world as a field *for* the use of tools."[68] "Technology is a mode of revealing," Heidegger writes in "The Question Concerning Technology," but unlike revealing as care—which "lets what presences come forth"—revealing as technology "has the character of a setting upon, in the sense of a challenging forth." Revealing as technology, *Dasein* is no longer open to the Being of beings but rather forces on beings an interpretation in accordance with their resource value; "a tract of land," for example, "is challenged in the hauling out of coal and ore [and] the earth now reveals itself as a coal mining district, the soil as a mineral deposit." Within this mode of revealing, "everywhere everything is ordered to stand by, to be immediately on hand, indeed to stand there just so that it may be on call for a further ordering"; beings are thus no longer revealed as unique manifestations of Being but are revealed as "stock" or "standing-reserve," which exist only for our purposes.[69] Relating to the world through "technology," we do not *discover significance* as "always already" part of the world but rather we *assign value.*

Highlighting the Cartesian assumptions underlying revealing as technology, Macquarrie describes it as fundamentally "calculative [thinking which] . . . objectifies and breaks up the whole, [which] is directed toward the handling and mastery of the things within the world, [and which] is concerned with beings, not with Being,"[70] and Bruce Foltz explains that it "discloses an imperative of total control and domination."[71] Indeed, as Heidegger sees it, the very core of revealing as technology is the Cartesian assumption: "One type of being, the human being, believes that all of Being exists for it."[72] By contrast, revealing as care hinges on the contrary claim that *Dasein* is one being among beings, albeit a unique one. Steiner summarizes the significance of this different starting point in shaping *Dasein*'s mode of revealing as either care or technology:

> For Descartes . . . the self becomes the hub of reality and relates to the world outside itself in an exploratory, necessarily exploitative way. . . . For Heidegger, on the contrary, the human person and self-consciousness are *not* the center, the assessors of existence. Man is only a privileged listener and respondent to existence. The vital relation to otherness is not, as for Cartesian and positivist rationalism, one of "grasping" and pragmatic use. It is a relation of audition. We are trying to "listen to the voice of Being." It is, or ought to be, a relation of extreme responsibility, custodianship, answerability to and for.[73]

Care as "letting be"
If our uniqueness as *Dasein* translates into environmental stewardship by positioning us as shepherds of Being, then our relating to the world through "care" also becomes significant in environmental philosophy. It is, after all,

through revealing as care that we allow the "self-manifesting of beings" and thus fulfill our role as shepherds; "for Heidegger," writes Westra, "shepherding means . . . allowing Being to be, while allowing ourselves to be what we truly are, there (*Da*), a part of Being ourselves, not a separate, different entity, treating of alien matter."[74] Thus any discussion of *Dasein* as care "must address the issue of man's relation to the earth"; and care emerges specifically as "Heidegger expounds the possibility of saving the earth and dwelling upon it . . . in contrast to the [anthropocentric] project of dominating the earth."[75]

In Heidegger's own words, "Mortals dwell in that they save the earth. . . . Saving does not only snatch something from a danger. To save really means to set something *free* unto its own essence."[76] Foltz expounds upon this central discussion of "saving" in Heidegger's work: "'to save' means not only to rescue, or else to lay aside and conserve, but also to preserve and protect as intact. To save something is to let it be and remain what it is, to deliver it over to its own possibilities."[77] And in this light, Stenstad notes the link between revealing as care and involvement with the beings at hand: "To care for things is to let them reveal themselves, refraining from the violence of forced disclosure. . . . This 'letting be' is not a drawing back or lack of engagement, nor a sense of detachment from 'things in general.' Dwelling means heeding and taking care of particular things . . . [and depends on] heedful attentiveness."[78]

Primarily, then, care for nonhuman beings is to take the form of letting beings be, or allowing the "self- manifesting of Being" in the natural world, and it is directed toward those beings that comprise our "world" by having a place of significance in it. Thus care for nonhuman beings is not to be understood in its more common meaning of "love," although Heidegger does also intend to capture this sense of the word as part of "letting be": "To embrace a thing or a person *in its essence* means to love it, to favor it."[79] The phrase "in its essence" suggests the interpretation of "letting be" as "letting its *being* come forth,"[80] and Zimmerman explains this as "letting beings manifest themselves with the least interference and the most cooperation."[81] "Letting be," like love, then, seems to have these two components: (1) allowing beings to unfold in their own Being without interfering or "challenging" them to Be in accordance with our needs, and (2) actively promoting their Being themselves by acting concernfully on their behalf, by preserving and protecting them.

Considering the first of these two aspects of care as "letting be," Westra highlights the importance of paying careful attention to beings in order to understand their Being, and thus she suggests that even the noninterference aspect is neither passive nor neglectful:

> This letting be is not an attitude of laissez-faire, in the sense of lack of concern or interest. If we say "let me be," we mean "leave me alone" or "don't concern yourself with me." Heidegger instead wants to understand . . . letting beings be in the sense of manifesting care, interest, concern, in order to understand what they truly are and to allow them to be just that. I don't let a man be what he truly is if I treat him as a pole, . . . nor do I treat a fish as what it is if I put it on the grass in the sun. The necessary prerequisite of "letting beings be" is to consider seriously what they are and to care enough about their being to allow it to be. . . . If we understand a fish as a fish, and a man as a man, we know truthfully what they are and know the truth of their being. . . . When our actions reflect our understanding they embody the essence of truth: we allow beings to be themselves through our understanding of what it is for them to be.[82]

Zimmerman similarly hints at the overlap of the noninterfering and protecting aspects of letting be: while "the best course of 'action' is to let beings be, to let them take care of themselves spontaneously in accord with their own natures, . . . [care as] 'dwelling' [also] means to cherish in the sense of preserving and caring for things."[83] And Heidegger himself implies that the second aspect of "saving"—"to seize hold of a thing threatened by ruin"—depends crucially upon the first—"to fetch something home into its essence."[84]

As his discussion of care confirms, Heidegger's project is not to give rules for what we should *do* but rather to encourage us to rethink how we should *be* in light of a truer understanding of what we *are*. Heidegger thus speaks of "finding what is fitting": "for man it is ever a question of finding what is fitting in his essence which corresponds to [his] destiny [as *Dasein*]; for in accord with this destiny man . . . has to guard the truth of Being [as] . . . the shepherd of Being."[85] Determining the specifics of care in a given situation, then, involves careful attention to nonhuman beings (in order to understand their Being), openness to encountering them as themselves rather than as what we want them to be (for us), and consideration of the "fitting" response in light of the call for both noninterference and protection. Given the dominance of revealing as technology, however, it is not at all uncommon for *Dasein* to betray "what is fitting in his essence," to deny our role as shepherd of being, to force beings to disclose themselves as resources, and thus to relate to the world as if it was a collection of things existing for our use. In the next section, then, let us examine the implications for environmental philosophy of understanding *Dasein* in terms of both revealing as care and revealing as technology.

"Care," "technology," and "use" of the natural world
Just as encountering objects in the world as present-to-hand things is secondary to encountering them as ready-to-hand (that is, it is not the way we
most fundamentally encounter them, when we are being truly human), so
too is the apprehension of nature as present-to-hand secondary to understanding it as ready-to-hand. Foltz clarifies the difference between these two
ways of understanding nature and confirms Heidegger's insistence on the
primacy of the latter:

> Prior to [the Cartesian knower's] distinction between facts and values, nature
> is more fundamentally revealed as something which is always already mean-
> ingful. . . . Nature is not principally disclosed as an indifferent given, but as "en-
> vironing nature"—as what surrounds us and is nearby from the beginning. This
> proximity, however, can be disregarded; only if this is the case can nature stand
> before us as merely present. "But when this happens," [he quotes from
> Heidegger] "the nature which 'stirs and strives,' which assails us and enthralls
> us as landscape, remains concealed. The botanist's plants are not the flowers of
> the hedgerow; the 'source' which the geographer establishes for a river is not
> the 'springhead in the dale.'"[86]

Zimmerman succinctly clarifies the essence of viewing nature as present-
to-hand: "The same dualism that reduces things to objects for consciousness
is at work in . . . reduc[ing] nature to raw material for humankind."[87]
Revealing as technology, we understand ourselves as Cartesian subjects, fundamentally separate from the natural world, and we "set-upon" nonhuman
beings, forcefully disclosing them as resources rather than as what they are
in and of and for themselves: when "Nature appears everywhere as the object of technology . . . the earth can show itself only as an object for as-
sault."[88] Revealing as care we understand ourselves as *Dasein*, concernfully
involved in the environment as part of our "world," and we interpret nonhuman beings as the manifestations of Being they are in and of and for themselves: "beings manifest themselves not merely as objects for human ends,
but in terms of their own intrinsic worth and within their own proper limits."[89] Westra clearly contrasts these two modes of revealing the natural
world: in sharp contradiction to the Cartesian subject's "technological way
of seeing—that enframing which somehow sees all being as revealing itself
as so many prepackaged goods, awaiting our pleasure, our present or future
use, . . . a [caring] shepherd does not view his flock as prepackaged cutlets:
he sees them as individual beings with needs to which he ought to be alert,
and a well-being with which he must be concerned."[90] Let us look now at

some specific instances of these two very different ways of understanding and relating to the natural world as they appear in Heidegger's work.

Heidegger writes in "The Question Concerning Technology" that the mode of revealing "that holds sway throughout [the modern world] does not unfold into a bringing forth [as in care] . . . [but rather] is a challenging" and that this revealing "concerns nature, above all." Nature is seen as "the chief storehouse of the standing energy reserve," and it "becomes a gigantic gasoline station, an energy source for modern technology and industry";[91] revealing as technology, we "put to nature the unreasonable demand that it supply energy which can be extracted and stored as such." Is Heidegger suggesting then, that revealing as care is incompatible with any use of the natural world? Or is he simply trying to distinguish between a way of interacting with and using nature that acknowledges its own Being *in and of and for itself* (and is thus appropriate for *Dasein*) and a way of using it that understands its Being only as *for us* (and is thus inappropriate for *Dasein*)? His examples in "The Question Concerning Technology" confirm that the latter is indeed his purpose; throughout this article Heidegger contrasts uses of the natural world, which reflect an interpretation of it as *for us*, with other types of use that acknowledge the Being of the natural world *in and of and for itself*, but nowhere does he label *use* itself inappropriate for *Dasein*. Let us look briefly at three of his examples.

1. *Wind power*. For his first example, Heidegger asks us to consider the windmill as a power source whose "sails do indeed turn in the wind [but] . . . are left entirely to the wind's blowing" in contrast to the coal-fired factory, which operates as a result of a chain of "challengings" of the natural world: "a tract of land is challenged in the hauling out of coal and ore . . . [and] the sun's warmth is challenged forth for heat, which in turn is ordered to deliver steam whose pressure turns the wheels that keep a factory running." One important difference between these two uses of the environment in producing power is captured by the application of the term "standing reserve" to the latter use of nature's energy (to denote the interpretation of "resources" that are gathered in order to be constantly available for use): "the windmill does not unlock energy from the air currents in order to store it" for additional uses in the future, but "the coal that has been hauled out . . . is being stored [and] is on call, ready to deliver the sun's warmth that is stored in it." Wind is an energy of nature that, in and of itself, turns things, so the windmill represents a use of the environment that respects its Being; the coal-fired plant, on the other hand, uses solar heat as an energy of nature by "challenging" the earth to produce

coal—by revealing the earth "as a coal mining district"—and then by challenging stored-up coal to produce heat, thus interpreting the environment not on its own terms but as a stockpile of energy existing for our use.

2. *Agriculture.* In his second example, Heidegger compares the use of land for growing food by peasant cultivation and by the modern agriculture industry. The farmer's cultivation "takes care of and maintains the field . . . [and] does not challenge the soil of the field"; "in sowing grain [the peasant] places seed in the keeping of the forces of growth and watches over its increase." Under the technological mode of revealing, however, "even the cultivation of the field has come under the grip of another kind of setting-in-order, which *sets upon* nature . . . in the sense of challenging it"; agriculture, which now has become "the mechanized food industry," depends on challenging the air to yield nitrogen for fertilizer and on challenging the soil through the application of fertilizer and pesticides and massive quantities of water in its quest to produce unnaturally high crop yields. In the peasant's garden, then, the natural interaction of soil, seed, air, and water is taken advantage of in order to cultivate crops for human uses, but this is in stark contrast to the control, manipulation, and artificialization of these same beings in the agriculture industry. And note that the image of the field itself appears different when it is the site of natural beings and forces being themselves than when it is the setting of artificial inputs and disregard for its Being: as technology, in the latter case, we have revealed the field as a resource for the production of human food, while as care, in the former case, we have revealed it as itself and made use of it accordingly.

3. *Water power.* For his third example, Heidegger considers the use of the Rhine river as a power supply:

> The hydroelectric plant is set into the current of the Rhine. It sets the Rhine to supplying its hydraulic pressure, which then sets the turbines turning. This turning sets those machines in motion whose thrust sets going the electric current for which the long-distance power station and its network of cables are set up to dispatch electricity. In the context of the interlocking processes pertaining to the orderly disposition of electrical energy, even the Rhine itself appears to be something at our command. The hydroelectric plant is not built into the Rhine River as was the old wooden bridge that joined bank with bank for hundreds of years. Rather, the river is dammed up into the power plant. What the river is now, namely, a water-power supplier, derives from the essence of the power station.

We might consider here as an alternative a water wheel, built into a river and taking advantage of the river's being itself; in this case, the water wheel takes its essence from the river (rather than the other way around) and leaves the essence of the river itself intact. Again we see that the problem does not lie in use of the river for human purposes but rather in revealing the river *as* a "water-power supplier," a being which exists *for us* and is thus given significance only in accordance with the instrumental value we place upon it, and thus in challenging it to "be" in such a way that meets our needs rather than allowing its Being to unfold in its own way (a way that we just happen to be able to make use of in meeting our needs).

And, finally, viewing nature as present-to-hand reinforces viewing each other in the same way—as resources—rendering revealing as technology doubly problematic and making clear the linkages between "saving the earth" and "saving ourselves." As Heidegger writes in "The Question Concerning Technology,"

> The current talk about human resources, about the supply of patients for a clinic, gives evidence of this. The forester who measures the felled timber in the woods and who to all appearances walks the forest path in the same way his grandfather did is today ordered by the industry that produces commercial woods, whether he knows it or not. He is made subordinate to the orderability of cellulose, which for its part is challenged forth by the need for paper, which is then delivered to newspapers and illustrated magazines [which], in their turn, set public opinion to swallowing what is printed.[92]

Revealing as technology threatens our Being in the same manner that it threatens that of other beings: we, too, become not unique manifestations of Being but resources, things, standing reserve.

Technology "drives out" care
What is the nature of the relationship between these two very different ways of relating to or revealing the world? Obviously, we do have access to both modes, but it is central to Heidegger's commentary on the modern age that revealing as technology has become much more prevalent than revealing as care. Heidegger claims, in fact, that part of what is distinctive about the technological attitude is this very tendency to displace and thus gradually eliminate revealing as care: "it drives out every other possibility of revealing."[93]

Let us consider Heidegger's discussion of curiosity as an example of this "driving out" of care by technology. Heidegger acknowledges two ways of observing entities in the world: marveling at them wonderingly and lingeringly with the intent to truly understand and appreciate their Being (associated with care), and looking superficially first at one entity and then at another "just in order to have known" (associated with technology). Attesting to the gradual displacement of care by technology, it is curiosity, the latter of these two modes of observation, that has become both the commonplace and the idealized mode (that is, curiosity is now seen as a virtue rather than as the deficient mode of perception it actually is). Curiosity concerns itself primarily with the surface of entities, with their superficial appearance, rather than with their very Being, and it overlooks the common in favor of the exotic; it fastens on that which "is not yet experienced or . . . is not an everyday experience [such that] . . . it does not dwell on . . . the constantly and immediately available things . . . but prefers characteristically to jump from one [new] thing to another."[94]

Although we may not like the implicitly negative tone of this description (a reaction that, in fact, confirms the reification of curiosity as a virtue), it does indeed seem accurate. And, in fact, we find it difficult to imagine a worthy alternative to observation as curiosity; after all, isn't the opposite of curiosity dullness, apathy, and lack of interest in anything? This (technological) mode of observing the world has thus largely come to displace the (caring) alternative: wonder, the consuming desire to understand as an end in itself, joy in the bewilderment that naturally follows deep attention to the unfolding of variety—the very experiences we often dismiss as "child-like" and "unscientific" or "unprofessional."

Perhaps the greatest danger inherent in revealing as technology, then, is this very tendency to conceal, not only the true Being of beings, but also the very possibility of revealing as care itself. Our ability to be "not-caring," to relate to the world through the attitude of technology, threatens not only those entities in the world (including ourselves) that we "challenge-forth" and objectify as "standing reserve" but also our own ability to relate to the world through care. Evernden addresses this twofold danger inherent in revealing as technology by suggesting the general form it takes in contemporary life:

> Not only do we lack the knowledge of how and for whom to care, but part of our new technological being hinges on the ability *not* to care, to remain dispassionate, unattached, and objective. This enables us to emphasize a kind of world, achieved through one mode of perception. But we also obscure that which is most important to us as living creatures. We gain explicit, barren fact and a world populated with useful things, but we lose all memory of why we are doing this at all, or why we *are*.[95]

This most fundamental threat to our Being is perhaps best epitomized by Cartesian anthropocentrism, our less than fully human mode of existence. "Saving ourselves" by "saving the earth" takes on an added dimension in Heideggerian environmental philosophy: "if we become frozen into a one-dimensional apprehension of beings, we will cease being human [*Dasein*]—for what is essential to our humanity is openness for novel and creative ways of apprehending what is."[96] Thus it is not merely that our physical survival is threatened by the environmental crisis, but also that the crisis—which itself is a product of our own deficient mode of Being—attests to our increasing inability to "be" in any other, more human, way. Not only does Heidegger's phenomenology have significant implications for environmental philosophy, but this extension of his thought in turn speaks to us even more immediately of our own existence as human beings in-the-world of other human and nonhuman beings.

Rediscovering "care" in a world of "technology"
We conclude this discussion of *Dasein*'s being-in-the-world as a field of care by exploring Heidegger's assessment of the prospects for our actually relating to the world through care given that the societal norms that shape how the world can "matter" to us are the very ones embodied in the not-caring of technology, the predominant Cartesian interpretation. It is important to note that, despite his emphasis on the significant role played by "the They" or "the One" in shaping our behavior, Heidegger does not argue that how *Dasein* can be-in-the-world never changes; rather, the specific understandings embodied in the public interpretation can change gradually over time (such that today it might be acceptable to do that which "one" never would have done in the past) as individuals come to recognize and (in a limited sense) transcend the pull of "the They" and thus, in effect, alter the content of what "They" tell us. Dreyfus explains this aspect of Heidegger's thought carefully and we quote him at length:

> [The fact that *Dasein* takes its everyday possibilities from "the One"] need not mean, however, that the roles, norms, etc., available to *Dasein* are fixed once and for all. New technological and social developments are constantly changing specific ways for *Dasein* to be. Nor does it mean that there is no room for an individual or political group to develop new possibilities, which could then become available to society. But it does mean that such "creativity" always takes place on a background of what *one* does—of *accepted* for-the-sake-of-whichs that cannot all be called into question at once. . . . This sociocultural background . . . can change gradually, as does a language, but never all at once and never as the result of the conscious decision of groups or individuals.[97]

Attesting to the importance to Heideggerian environmental philosophy of displacing the dominant mode of revealing as technology, Evernden suggests that "there is no possibility of an environmental ethic . . . in a society dominated by the technological vision of the world";[98] and, along the lines suggested by our earlier discussion of the role of the public interpretation, Reed adds the Heideggerian conclusion that "we hold the world hostage to whatever it is we think we are destined for."[99]

As specific examples of this process, consider that it was not so long ago that "one" would never allow one's bare skin to be seen in public or that "one" would never demonstrate affection in public, inhibitions that are now laughingly dismissed as overly self-conscious in the former case and prudish in the latter. And, of course, the same dynamic operates in the reverse direction: in today's world "they" look askance at a couple with eight or ten children, a practice that in former times evoked admiration. Or, to take an example of great ethical and political significance, "they" used to say that no proper Caucasian gentleman would perform manual labor when naturally inferior Negroes could so easily be purchased and enslaved to do it for him, a prejudice which "they" would go to great lengths to deny today.

But how do such changes come about, and is there any reason to hope that "the They's" predominantly and increasingly technological attitude will gave way to an attitude of care, that the public interpretation will change sufficiently to allow us to see the care alternative as a real option that might save ourselves and our world? Considering these very questions himself, Heidegger quotes Hölderlin: "But where danger is, grows / The saving power also";[100] and, suggesting that the saving power increases along with the danger (the increasing displacement of care by technology), he hints at the vehicle of our salvation: "We are . . . summoned to hope in the growing light of the saving power. How can this happen? Here and now and in the humble things, that we may foster the saving power in its increase."[101]

We will take a brief look here at three types of "humble things" Heidegger deems as having "saving power": marginal practices, art, and reflection. Each of these, in its own way, opens new ways of understanding and responding to our world, beyond what is embodied in the public interpretation, beyond what "They" tell us is appropriate.

1. *Marginal practices.* Marginal practices are those norms and behaviors that "were once central . . . but have now become rare and therefore are not what one normally does"; they are not condemned by "the They," but they are nevertheless seen as "old-fashioned, trivial, or meaningless."[102] Writing personal letters, bringing flowers and candy to the door on a date, and taking the family on Sunday afternoon drives are examples of practices that "one"

used to engage in regularly but that have become marginalized and uncommon. Thus such marginal practices as growing one's own food organically, walking instead of driving the car, and learning the constellations at night rather than going to the theater for entertainment "provide a basis for resisting the technological understanding of being": they offer "new" (actually marginalized old) ways of understanding and responding to our world, ways that do not reinforce Cartesian assumptions but rather remind us of other, non-"technological," ways of being.

According to Dreyfus, "Heidegger holds that at any given stage of history certain marginal practices will be especially relevant, so that in taking them up . . . *Dasein* can define the current issue for itself and its generation. . . . Ecology, for example, might be the issue for our generation, requiring . . . adapting past practices of preserving and respecting nature."[103] Clearly recycling, for example, is an instance of the current generation having picked up the marginal practice of carefully reusing goods—which, ten years ago, at least, "They" dismissed as an old-fashioned eccentricity in a proudly "disposable" society—and thus having gradually altered the public interpretation in the direction of care (today, of course, "one" would never simply throw away an aluminum can or pour one's used motor oil in a ditch). Wearing fur might be another example of an environmental ethics issue on which "They" seem to be gradually changing their minds and moving in the direction of revealing as care: owning fur coats, of course, was completely uncontroversial fifty years ago and, in fact, was the premier symbol of wealth and social status, whereas today the practice is largely frowned upon as ignorant, callous, and cruel.

2. *Art.* Heidegger suggests that revealing as care "lays claim to the arts most primally, so that they for their part may expressly foster the growth of the saving power";[104] art, in other words, is (or can be, for it too can be "technologized"—as in the deliberate attempt to create works of art in accordance with current standards of aesthetics) uniquely expressive of the Being of beings and is uniquely suited to engender wonder and the appreciation of Being. In a classic example, Vincent Van Gogh's painting *A Pair of Shoes*, we encounter the shoes not as "things" to be measured, evaluated, and assigned a value but rather as "a part of somebody's world"; we "reveal" them "in their context," appreciating that "they have a history, and to somebody they are very important," and realizing that "someone lives and works in them, and every scratch and tatter is an evidence of someone's whole existence, the entire world of a peasant farmer."[105] We might also note as an example the

increasingly popular murals ("Whale Walls") of Wyland. His large-scale paintings are painstakingly crafted to capture the true Being of whales—in their natural habitat, in the midst of their family groups, and in all their intelligence, playfulness, size, and friendliness. Wyland swims with whales and dolphins in order to understand them in their own world and to communicate with them, and, echoing Heidegger's understanding of the role of art in reminding us of Being, he has said of his work that "It is my job now to bring people to the side of the whale."[106] Art that truly captures the essence of a being, then, is both a product of and a vehicle for the experience of revealing as care.

3. *Reflection.* And, finally, Heidegger suggests rethinking and reflection as a third "humble thing" with the "saving power" of calling us back to our Being as care. "The coming to presence of technology harbors in itself what we least suspect, the possible upsurgence of the saving power . . . *provided that we, for our part, begin to pay heed to the essence of technology,*" Heidegger writes in "The Question Concerning Technology." And he expands on the role thus played by our reflecting on the essence of technology:

> Everything, then, depends upon this: that we ponder this arising and that we, recollecting, watch over it. How can this happen? Above all through our catching sight of what comes to presence in technology, instead of merely gaping at the technological. . . . We look into the danger and see the growth of the saving power. Through this we are not yet saved . . . [but] we may foster the saving power in its increase . . . [by] holding always before our eyes the extreme danger[:] That the coming to presence of technology threatens revealing . . . with the possibility that all revealing will be consumed in ordering and that everything will present itself only [as] . . . standing-reserve. Human activity can never directly counter this danger. Human achievement alone can never banish it. But human reflection can . . . questioningly ponder the essence of technology.[107]

Thus, careful consideration of the essence of technology (as a mode of revealing) must take the place of our current uncritical fascination with ("gaping at") the "toys" and tools of modern life; we must first of all become more aware of and sensitive to the danger we face in becoming increasingly technological. Through such reflection we develop "a serene openness to a possible change in our understanding of being."[108] Reflecting and rethinking, *Dasein*, "while using technological devices, stands outside [the] nihilistic technological understanding of being . . . into which everyone is socialized . . . [and] is thus open to other sorts of practices that still survive, and to a new understanding of reality, should one be given us by a new cultural paradigm."[109] And it may well be that the entire field of environmental philosophy is an effort to reflect on the danger posed by revealing as technology and to rethink the assumptions about ourselves and our world that have led us to the current "crisis" situation. As we have seen throughout this text, environmental ethicists are, almost by definition, committed to uncovering the roots of the current environmental crisis in the western worldview, and certainly they are encouraging us to think a bit more critically about the nature of technological progress and to reconsider the role we have assumed for ourselves in the natural world. Having become consciously aware of the historical development of the technological attitude and of the role of the public interpretation in reinforcing this attitude, we are more questioning and consequently more receptive to the "saving power" of all these "humble things."

Heidegger suggests that "the closer we come to the danger, the more brightly do the ways into the saving power begin to shine and the more questioning we become."[110] As these several examples confirm, we do indeed seem to be in the process of gradually altering the public interpretation of acceptable interactions with nonhuman beings and the natural world in the direction of revealing as care. In so doing, Stenstad concludes, "We move from disconnection to connection, from a refusal to hear to gathered attentiveness, from our panic-stricken obsession with control (and the violence which all too often accompanies it) toward the strength and wisdom to care for things in accord with how they show themselves to us."[111]

Reconsidering Environmental Ethics Issues

We have seen that the question at the heart of Heideggerian environmental philosophy is not whether the environment and nonhuman beings can be used for human purposes but rather whether current uses reflect the appropriate respect for beings *in and of and for themselves.* We might fruitfully consider in this light some of the issues involving our use of individual nonhumans examined in previous chapters; in so doing we will at the very least gain some practice in thinking from this perspective, and this experience will enhance our understanding even if we are unable to pin down all the specifics (which, we should note, would not be Heidegger's goal in any case).

Is it appropriate for *Dasein* to eat animals and animal products? Note, first of all, that we are not inquiring regarding the moral standing of nonhuman animals or the legitimate balancing of rights and interests; instead, our question concerns the compatibility of care and the consumption of those beings we are to "let be" in and of and for themselves. Although it initially seems like an obvious contradiction to speak of eating a being whose Being we respect, we must recall that care does not require nonuse. The question may be less whether meat-eating is appropriate than what understanding of ourselves and other beings renders it either appropriate or inappropriate. Factory farming—the raising and processing of animals for food, which involves restricted freedom of movement, manipulation of breeding cycles, artificialization of behavior, and other "challenging" practices—clearly seems to be on par with the coal-fired power plant and the damming of the Rhine: in no way does this industry allow beings to unfold in and of and for themselves, and, in fact, it is the very institutionalization of technological revealing, revealing animals as resources that exist for us. Does this imply, however, that all cases of animal consumption embody a similar disregard for their Being? It may well be that raising free-range chickens and cows—gathering eggs, milking, and even eating the animals when they die—is compatible with *Dasein*'s being as care; such a practice does, after all, appear to allow the Being of the cows and chickens to unfold without manipulation or control, and it appears to avoid the Cartesian characterization of the animals as present-to-hand things. Underlying such appearances, however, is the crucial distinction: whether we understand the chickens and pigs to be mere resources (for us) or independent manifestations of Being (in and of and for themselves) whose Being we just happen to be able to profit from.

And, of course, the same distinction is central to the question of recreational "uses" of nonhuman beings. The recent upsurge in activities such as whale-watching and wildlife photography seems more in line with an understanding of ourselves as "care" than does, for example, sport hunting. How could shooting a deer or a quail not be a violation of the very Being we as *Dasein* are to care for? Here again, however, we must consider the way in which the animal is revealed by such a practice: does the responsible, sportsman-like tracking and shooting of a deer reflect appreciation for the animal's unique Being as a fellow being in the ongoing cycle of life and death, or does it embody an understanding of deer as "stock," as a resource to be managed for human purposes? There may be no clear answer to this question, and doubtless many hunters would argue for the former interpretation, whereas many people who oppose the practice would argue just as intently that sport hunting is by definition resource exploitation. Is the recreational alternative embodied in whale-watching and wildlife photography any more amenable to clear-cut analysis, however? We can at least acknowledge that, just as women (and men) frequently experience exploitation that involves no physical harm, being perceived as sex objects, for example, so too might whale-watching be seen as an objectifying and exploitative although physically harmless form of recreation. Visiting Denali National Park while in Alaska conducting this research, for example, I frequently heard expressions of annoyance that the moose, caribou, and bear tourists had come to photograph would not come near enough to be seen clearly; surely such an approach to wildlife photography fails to respect the Being of beings and treats animals as "standing reserve," on call to be observed and photographed whenever we desire. It seems that, from a Heideggerian perspective, such an attitude has a great deal in common with damming the Rhine and factory farming even though it lacks the invasive, disruptive physical consequences of these more obviously not-caring practices. Again, the relevant factor is less the actual behavior (or "use") and more the underlying interpretation of the Being of other beings. Certainly whale-watching can be a form of recreation that reveals as care and is thus appropriate for *Dasein*—when it is driven by the desire to experience whales as independent beings in their own world and by appreciation of their unique manifestation of Being or, in short, by wonder; but, as seems to be the case with just about any human interaction with the natural world, it also holds the potential for the inappropriate revealing as technology (or, more precisely, *we* hold the potential for both forms of revealing and thus we, in our relationship to Being, and not the situation itself or the types of beings involved, are determinant of appropriateness and inappropriateness).

This last parenthetical distinction makes clear the crucial difference between this approach to environmental philosophy and that of "environmental ethics" per se, especially as embodied in extensionism. It relies neither on assessment of the "values" of nonhumans or the natural world in general nor on consideration of nonhuman rights and interests, nor, in fact, on any characteristics of or consequences for the nonhuman "other" at all, but rather finds the morally relevant consideration in ourselves, in our way of being; from a phenomenological perspective, the appropriateness or inappropriateness of any given interaction with other beings "depends on the stance we take towards the world, and not on the things we encounter."[112] Our being what we are, *Dasein,* depends crucially on letting nonhuman beings be what they are, on relating to them through care; and this is the basis for distinguishing appropriate and inappropriate interactions with nonhumans and the natural world. As Zimmerman paraphrases Heidegger, "only Being can impose appropriate limits on humankind"; and this means that "we must learn to behave in accordance with who we really *are*, in accordance with our own Being."[113] Thus, here the project of environmental philosophy involves not ethical theory but establishing a new "*ethos*": "a new paradigm for understanding what we and other beings *are*, . . . a profound understanding of and respect for the Being of *all* beings."[114] Foltz captures the essential difference between the approach of environmental ethics and that of Heideggerian "*ethos*":

> What is at stake here is not just our eagerness or hesitancy to fell a tree, but whether that tree is to be [a being] or merely a supply (either potential or actual) or BTUs. Nor is this simply a question of our *attitude* toward the tree, but of what it is allowed to *be*. Inhabitation or dwelling, "staying with things," is necessarily an act of tending and attending which grants things the leeway to disclose themselves and endure; an inhabitation in this sense is a genuine "caretaker." And a relation to the natural environment based upon such heedful inhabitation is in itself a recovery of the original basis for an environmental ethic: a "familiar abode" or *ethos* "upon the earth and under the sky."[115]

Conclusion

In this chapter we have examined the phenomenological tradition in general, especially by contrasting it with the tradition of Cartesian rationalism, and we have explored Heidegger's thought in this light, paying close attention to the unique understanding of ourselves it offers: ourselves as *Dasein*, absorbed in the world, distinctly aware of Being and not-being, and thus essentially relating to the world appreciatively and respectfully through "care." And we have explored the relevance of Heidegger's thought for environmental philosophy and have come to appreciate the ways in which familiar "environmental ethics" issues are recast and rethought in this light. We are once again prepared to reduce this tradition to its core aspects, noting in the process the "extension" of these elements specifically into the realm of human–environment interactions.

An Ontological Rather Than an Ethical Orientation

Heidegger's work is not a body of ethical theory but rather an investigation in the realm of ontology; to the extent that it has a normative tone, it is concerned with "how we should *be*" not with "what we should *do*," and the guidelines it offers for "how we should be" come from consideration of what we *are*: beings uniquely aware of Being and thus under the special responsibility as "shepherds of Being" to respect (care for) the Being of other beings. The situation before us and the attitude toward Being we act from are determinant of the appropriateness or inappropriateness of our Being in that situation, and we are called on to respond to the world in which we are fundamentally involved in a way that is "fitting." Dreyfus quotes Aron Gurwitsch, another Heidegger commentator who distinguishes this approach—*Dasein* trying to respond appropriately to the world in which he is "always already" immersed—from that embodied in ethical deliberation—the moral agent adopting the moral point of view:

> We find ourselves in a situation and are interwoven with it, encompassed by it indeed just "absorbed" into it. . . . What is imposed on us to do is not determined by us as someone standing outside the situation simply looking on at it; what occurs and is imposed are rather prescribed by the situation and its own structure; and we do more and greater justice to it the more we let ourselves be guided by it.[116]

Rather than viewing the "other," or rather his characteristics (his rights, his suffering), as the morally relevant determinant of right or wrong action, Heidegger's ontological approach places the relevance on the "self" and thus judges any given behavior appropriate or inappropriate depending on its conformity with our true Being. Wonder at the beings with whom we are absorbedly involved rather than the appeal to rules is central.

Heideggerian environmental philosophy thus does not aim to develop and defend a set of guidelines for our interactions with nonhumans and with the natural world at large but rather simply offers a new way to assess the appropriateness or inappropriateness of these interactions: consideration of our Being as *Dasein*. The project here is not to reconsider the question of moral standing and then to apply the standard rules of ethical theory to our interactions with the new nonhuman members of an expanded moral community; rather, we are to ask ourselves whether our relationship with the natural world is in accordance with what we really *are*: shepherds of Being whose true essence as "care" lies in "revealing" other beings as they are *in and of and for themselves*, not as resources *for us*. As we noted in our attempt to extend Heidegger's thought to such environmental ethics questions as factory farming and sport hunting, the relevant consideration lies in *our Being* rather than in *their characteristics*. Neither animal suffering or rights nor the attribution of environmental values serves as the determinant of right or wrong action or the guideline for our behavior. The situation and our Being in it, whether or not our response is "fitting" to the situation and to our nature as *Dasein*, determine the appropriateness of our thoughts of and actions toward nonhumans. The experience of wonder at Being, our own Being as well as that manifested by nonhumans, rather than the appeal to rules is the core of Heideggerian environmental philosophy.

A Replacement of Cartesian Presuppositions with an Understanding of Reality as We Actually Experience It

Heideggerian phenomenology, like feminist theory's care tradition, critiques the Cartesian worldview and develops an alternative based not on presuppositions about the world and our place in it but rather on an experiential understanding of reality. Thus, *Dasein* is not primarily a detached knower but an involved doer; he is not primarily a deliberator but an active participant; and this model of the human being is derived from careful attention to our everyday existence, not from a priori, abstract theorizing. At the heart of this phenomenological approach is a call for the renewal of awareness, appreciation, and wonder at every facet of existence, an experience not only rendered meaningless but also made virtually impossible by Cartesian assumptions.

Heideggerian environmental philosophy dismisses Cartesian assumptions about our place in nature and seeks to develop a new understanding of human–nonhuman interaction in light of actual experience. Thus the Cartesian presupposition according to which our unique rationality sets us apart from and in the position to exploit the natural world is replaced with a different understanding of our place in the world: we are indeed unique, but our uniqueness lies in our awareness of Being and not-Being, and our role is thus not one of mastery over beings but one of responsibility for Being itself. Cartesian anthropocentrism thus gives way to the Heideggerian image of *Dasein* as the shepherd of Being. Similarly, our tendency to relate to the world as if we were indeed Cartesian subjects detached from the world of "things" that exist for us is viewed from this perspective as a deficient mode of Being: it is what we are when we are not true to our Being as *Dasein*. The natural world is thus not primarily a present-to-hand thing to be known and characterized in objective terms. Heideggerian environmental philosophy, unlike the Cartesian worldview, acknowledges as valid our intimate relationships with and subjective experiences of the natural world in which we are fundamentally absorbed. The human realm is no longer defined in opposition to the realm of nature but rather the natural world, and nonhuman beings are seen to be integral parts of our experiential world.

A Characterization of the Human Being as Fundamentally in-the-World

Dasein, unlike the detached Cartesian subject who surveys the world from the position of a would-be knower and manipulator and who is only contingently related to others, is most fundamentally in-the-world; he does not merely survey a collection of things in the world "out there" but rather encounters objects and other people as "always already" related to his own Being. And this Being-in-the-world is best modeled as a field, an expanding and contracting experiential region that is composed of significance, rather than as a physical entity spatially "in" a setting of other separate entities. Heidegger's term "de-severance" accounts for the experiential, non-spatial nature of our relationships with other beings who constitute our world. Corollary to this non-Cartesian understanding of the human being as *Dasein* is a focus on our Being in terms of the "activities we pursue and the things we take care," on concerned participation rather than detached deliberation.

The natural world at large and nonhumans in particular are not something out there, removed from our existence and only contingently related to us as a thing to be exploited but are rather always already related to us. Experiencing ourselves as fields or regions of care, we encounter nonhuman beings concernfully in the course of our active involvement with the natural world. Along these lines, Evernden makes a distinction between the natural environment that the Cartesian subject is (spatially) "in" and the natural world that *Dasein* is (experientially) "in," and in doing so he suggests that, for some of us, some nonhuman beings—his example is the two hundred thousand species of beetles—are not part of the world of significance: "they do not *function* in the lives of most humans [and] this is not to say that they could not, or should not, but that they do not."[117] Different beings, then, may at different times be part of or excluded from any given person's world; and, of course, the more we tend to exclude other beings, by virtue of our failure to appreciate their Being and their significance, the more our Being approaches that of the Cartesian subject, detached from the environment rather than involved in the natural world. Evernden, we recall, extends the model of the human being as a field of concern in-the-world to account for the environmentalist's defense of the natural world: as a field of care we "hear our name being called" when nonhuman beings which have a place of significance in our world are threatened and we react out of our Being as care.

A Recognition of Two Fundamental "Attitudes" or Ways of Interpreting the World: Care and Technology

Throughout Heidegger's work, of course, one of the central themes is the recognition of and distinction between care and technology as two modes of our Being-in-the-world. It is in discovering beings in the significance of their Being, in "disclosing" them in their own Being (in and of and for themselves), that Dasein relates to the world through care; relating to it as a Cartesian subject, assigning value to beings "disclosed" in terms of their exploitability (for us), is also a possibility for the human being, but this revealing as technology is a secondary and thus less authentic, less fully human mode of Being. The alternative of relating to the world through the "attitude" of technology is, in fact, a product of our misunderstanding ourselves as the "thinking thing" who is fundamentally a user of "nonthinking things" which in turn have only instrumental value. Revealing as care, by contrast, is *Dasein*'s "ownmost and most proper" way of Being; it both depends on and engenders attentiveness to beings, the appreciation that comes from intimate knowledge pursued in a spirit of wonder, and a sense of responsibility for the Being of beings.

These two modes of "revealing" are most clearly demonstrated in our interactions with the natural world. While the Cartesian worldview can only distinguish between sustainable and unsustainable resource exploitation, Heidegger's environmental phenomenology uncovers a deeper explanation of environmental crisis: our less than fully *Dasein* way of Being or our deficient mode of revealing the natural world as exploitable "standing reserve" existing solely for our use. Heidegger's image of the Earth as a "gigantic gasoline station" captures the interpretation of the natural world the technological attitude yields. When, however, we are true to our Being as *Dasein,* we reveal the natural world through care; other beings are seen as the unique manifestations of Being they are rather than as resources. Care need not imply nonuse of the natural world (as we saw in Heidegger's examples of the windmill, the cultivation of crops, and the waterwheel and in our own extension of this model to the questions of diet and recreational use), but it does require careful consideration of what such use suggests about the way we thus reveal the natural world. We have seen that care for nonhuman beings involves attentiveness to them, so that we can come to truly and wonderingly understand their Being, and "letting be": allowing their Being to unfold in its own way, without interfering or "challenging" it forth in accordance with our needs, and sometimes cooperating with and protecting this unfolding of their Being. To summarize, Heidegger's distinction between revealing as care and revealing as technology is paradigmatically expressed in our dealings with the natural world and embodies a very different perspective (from that of environmental ethics) on appropriate use of the natural world.

An Acknowledgment of the Role of Public Interpretation in Shaping Behavior

The dominant understanding of what it means to be human and of how the world is valued, however, is largely Cartesian in nature, and it is another of Heidegger's primary points that both our understanding of ourselves and our behavior are shaped by this public interpretation and thus diverge, increasingly, from our "ownmost and most proper" way of Being. We take our behavioral guidelines from what "They" tell us about ourselves and our world—we tend to do what "one" does and to avoid what "one" does not do—and in the modern era "They" increasingly tell us that we are subjects in a world of objects that are to be exploited for our purposes. The "technological" attitude thus predominates because it is legitimated by the (Cartesian) public interpretation from which we take our norms and standards and through which we learn our place in the world. Heidegger's task, then, and the one that he encourages us to participate in, involves exposing the mistaken roots of the public interpretation so that it can gradually be displaced by an alternative that acknowledges the possibilities of our Being as care. Overcoming this public reinforcement of the attitude of technology is doubly important given the tendency of this mode of revealing to eliminate the more authentic mode of revealing as care and thus to diminish our Being *Dasein.*

By the same dynamic, "we see in Nature what we have been taught to look for [and] feel what we have been prepared to feel."[118] And, of course, "They" tell us that the natural world is an energy source, that animals exist for our consumption and amusement, and that we have dominion over the Earth; and we, therefore, think and behave accordingly, viewing the natural world as a resource and ourselves as its legitimate exploiters. Environmental policy thus becomes the realm for rational resource management in which nonscientific, nonobjective experiential understandings of our interaction with the natural world are invalidated and dismissed. Environmental ethics theorizing itself, in fact, is largely structured by the public interpretation and thus from this perspective poses no fundamental challenge to the Cartesian worldview but rather works within that framework; Evernden makes just this point about environmental ethics, claiming that "it also arises within a culture, and can only do, or even aim to do, what cultural assumptions reveal as possible."[119] Environmental phenomenology, on the other hand, can both account for and contribute to the gradual alteration of the public interpretation away from revealing as technology and toward revealing as care. We have seen in the examples of recycling and fur wearing that the content of the public interpretation is indeed gradually changing with respect to appropriate interactions with nonhumans and the natural world; and, indeed, from the perspective of Heideggerian phenomenology we can also recognize the field of environmental ethics as part of this gradual alteration of the public interpretation insofar as it is committed to rethinking and reflecting on our place in the natural world (recall that this is one of Heidegger's three "humble things" with the "saving power" of overcoming the public interpretation and returning us to our true Being as care). A true environmental ethic—or, rather, *ethos*—of course, hinges on just this alteration of the public interpretation of our role in the natural world; and "in encouraging wonder," Heideggerian environmental philosophy "simply prepare[s] the ground for the public germination of an idea"[120]—experienced if not often expressed—of ourselves as "shepherds" rather than masters, of nonhuman beings as unique manifestations of Being rather than resources, and of the relationship between us as one of wonder, appreciation, and care rather than detachment and exploitation. The move from understanding ourselves as Cartesian subjects to understanding ourselves as *Dasein* and the transition from revealing as technology to revealing as care not only embody Heidegger's prescriptions for more appropriate human-nonhuman relations but also constitute the "turn in the Western tradition [which] will release us from our present alienated condition."[121]

It was as if we had resolved to demonstrate to the rest of creation that . . . when our best instincts are fully engaged, man no longer is the planet's most treacherous animal.

But it also raised troubling questions about the human proclivity either to pretend that animals are more like people than they are or to treat them as mere commodities.

Chapter 7 Whale Rescue Story III

For the third and final time, we turn now to a recounting of the whale rescue story from the unique perspective offered by the conceptual lens currently under consideration: how will our imaginary Heideggerian ethicist answer our four central questions, and what new light will be shed on the event's unfolding?

In the previous versions of the story (Chapters 3 and 5), we saw the influence of the core assumptions of extensionism and ecofeminism. In Story I, the imaginary ethicist justified the effort on the grounds of principle and duty; opposition to the rescue was largely based on frustration with the sentimental, unreasoning responses that were thought to drive it; and the ethical dilemma was cast in terms of prioritizing competing interests (human or whale, individual or system) and, accordingly, was understood in largely abstract terms (whales, not these three whales). Story II, however, explicitly reflected a different set of assumptions. Emotion, empathy, and compassion were acknowledged as legitimate motivators; the responsibility to help was grounded in the nature of relationships with the whales rather than in an expanded moral community; and at least part of the opposition to the effort was seen to be based in the judgment that it actually constituted harm.

Having seen our previous two conceptual lenses in operation, we should be able to anticipate the unique shape Story III will take (although, again, readers should note that the following account is but one scenario among several possible interpretations open to this perspective). We might expect our ethicist to cast the dilemma in terms of the two aspects of "letting be," noninterference and preservation, and to justify the rescue effort in terms of what is "fitting" for human beings given our responsibility as "shepherds." We might expect less of an emphasis on deliberation and more on absorbed action, we might assume a significant role for what "They" say regarding whales, and we might expect respect for Being rather than rules or principles to be the central concept.

So once again, let us pose our four central questions, noting both the correspondence between this version and the philosophical framework outlined in the preceding chapter and the uniqueness of this third recounting.

Why was the rescue effort the appropriate course of action?

(1) We related to the whales through care, our ownmost and most proper way of being.
"It's the human in us." As human beings, we responded to the plight of the whales out of the "care instinct we have, . . . [the] instinct to save . . . lives . . . [and to] be kind to the animal world and to the environment. . . ." "Protecting species . . . and rescuing individual animals . . . are [part of the] 'humane stewardship of life' . . . [which is our] responsibility." "People want to have something to care about; they want to have something to feel genuine about, and this was a very powerful image of something to care about outside of ourselves." One observer saw the rescue's confirmation that our humanity lies in care as its most significant element:

> The deeper attraction of the story may have lain in what it showed about the people attracted to it in the first place. The freeing of the whales, whatever else it demonstrated, was a basic fable in human generosity. Nothing called for all that human labor, after all, but an impulse to preserve a couple of lives and to let them take their course. . . . [Thus, the rescue demonstrated] the observable fact that people are driven as naturally to preserve as to destroy.

"It is a human trait" that "[we] can't just stand by [when] there are these three helpless creatures" and that "if we don't do something about it . . . we'll feel a part of us died." "That's what makes us human, that we have those impulses . . . to rescue and to save and to want to see things work out well and to feel like we're personally caring, . . . and when we act on them . . . that makes us better than human." "It is rare in anybody's lifetime when you have an opportunity to get involved and save a piece of your environment, to save another living organism," so of course we would take advantage of this "opportunity to be mindful"; "we want to feel good about being human, . . . to transcend the ordinary in humans, . . . to be connected to each other and to the world, . . . [so] we responded on a deep level [to the whales]."

(2) We acknowledged that they and we are equally beings in and of and for ourselves.

An important aspect of our being truly human and relating to the world through care involves denying Cartesian anthropocentrism and acknowledging instead that other beings exist in and of and for themselves just as we do. "We are all beings on this earth"; "these creatures share our environment" and "have an inherent worth just like we do," and "we are just another link in the chain, humbling as that fact may be." In part, then, the whale rescue was the appropriate course of action because it acknowledged that "they are not things, [that a whale is] a being in and of itself." In one commentator's assessment of the rescue: "The sense that it does make is simple enough. We finally know human life is tied to the life of plants and animals. We know, better than we did, that we are part of nature, not disembodied from it and not enthroned above it. And as we value our own life, we increasingly value life in general."

(3) We responded to impending non-Being.

Given our unique awareness of Being and not-Being, we as human beings are responsive to impending non-Being. The whales were "part of a vanishing wilderness and we hate to add to the demise and therefore try to preserve as much as possible." Several young onlookers expressed this awareness of and response to the potential extinction of whales in their letters urging the rescue forward: "I hope you all keep going and don't give up because the Gray whale is almost extinct and I want them to live"; and "It would be really sad if the California Gray Whales were extinct. They're already endangered"; and, again, "I think it is very nice what you are doing. I couldn't imagine the world without our whales." Frequent comments during the news coverage to the effect that "if help doesn't come soon, there's no question that they can't survive" served as a call to many of us to respond, just as, upon the death of Kannick, "the nation mourned" and the rescue effort took on a renewed urgency.

(4) We promoted the unfolding of the Being of other beings.
Being true to our being as "shepherds," caring for and respecting the Being of other beings, is primarily a matter of promoting the unfolding of their Being, sometimes through noninterference and sometimes through active protection. In this light, one commentator on the whale rescue wrote:

> We should do two things. We should make sure we do not interfere in any large or irreversible way with the cycles of life established through the eons in nature. We should also make sure we develop our own natural compassion.
>
> We should not let spiders be sprayed to extinction. But we may interfere to free a small bug trapped in a web. We may not allow fish to be poisoned at random through careless use of chemicals, but we may on occasion fish out a waterlogged bug from the fish pool that would otherwise be devoured.
>
> Our actions may not always make sense—what kind of sense does it make to take a spider's food away or keep a fish from devouring a honeybee, while urging that spiders and fishes get plenty to eat, in an environment of clean air and unpolluted water?
>
> It makes sense all the same to get a cat out of a tree, to rescue a dog from a swimming pool, to send costly equipment to rescue whales.

While respecting the Being of other beings may sometimes require standing aside and letting them take care of themselves in accordance with their own natures, it may also require "reach[ing] out to help them in times of great emergency"; often, respecting an animal and behaving in such a way as to further the unfolding of his being means that "when need be, it is protected, as is the case with the whales."

In this case, it was obvious to anyone who paid attention that the whales' Being—not just their lives, but their very essence—was constrained and thwarted by their entrapment. Media coverage described them repeatedly as "lonely and weeping," as "battered and weakened by their constant battle to surface for air," as "pathetic," and as "grasping at the edge of life [and] becoming weaker and weaker by the hour"; and Kannick especially was observed to be "tired . . . and weak . . . [and to] come up to rest himself against the side [of the air hole]." Fewer images could reflect further deviation from the true being of a whale: whales are supposed to be social and communicative, to breach the surface of the ocean in joyful abandon, to be among the strongest and most graceful of animals, to surface and breathe effortlessly, and to be virtually tireless "perpetual motion machines." Whales "[have] this wonderful free-ranging life, . . . spend[ing] their winters in Hawaii and their summers cruising around . . . Alaska completely free" and so "for most of the world, they just couldn't fathom these whales trapped in the ice." Whales are "supposed to be free," not "captives of their environment," and the media's juxtaposition of their entrapment and the "frolicking" of bowhead whales in the nearby leads made it clear that respect for their Being required "get[ting] the whales out of their horid [*sic*] cage."

And along the way to that ultimate goal, it also required modifying the conditions of their entrapment such that their true being could flourish to the greatest extent possible in the interim: air holes in the ice needed to be modified into longer channels so that the whales could surface in their normal arching fashion rather than in the constrained spyhops which were all the original small holes permitted. For many participants, in fact, the moment the whales entered these channels and were able to surface and dive normally "made it worth it all": "a whale surfaced and did something no one had seen these grays do. [He] rolled in a graceful arch, blew spray into the air, and continued a rolling swim which took [him] back under the water. A loud 'hey, hey, hey,' interspersed with sounds mimicking various sea mammals, rose into the air. . . . 'Finally, they look like whales!' the shout of a woman was heard." Kannick's death was even more tragic in this light, however: "It was like, wait a minute he can't die now just as we've got these holes open; it was the timing which was so sad, because we'd just finally got it so they'd be more comfortable."

"Looking in their eyes, you could see they wanted to be free," and it was clear that respect for the whales' Being, in this situation, required at least trying to return them to the open ocean and to the normal course of their lives. "What [was] the alternative? Watch them die there or let them die their own death out there in the ocean where they are at home and there's always that chance that they will live."

(5) We acted concernfully and absorbedly in our world.
Many of the participants "had a real reaction . . . related to their . . . own feelings of connection [with the whales]." "If it is in your space, there's something calling you to act," and the trapped whales were indeed in our "space"—our world—whether Inupiat, Alaskan, or citizen of the Lower 48 or even of another country. The Inupiat Eskimos "have always lived with these animals," and in the words of a spokesperson, "It's just that we care

for the environmental part of our culture. . . . We care for these animals, we depend on these animals. . . . We have to care for what passes by here, not to defeat it." And similarly for the more distant members of the North Slope and Alaskan neighborhood, "stewardship . . . includes taking care of the areas that we have, the areas that we hunt and fish, travel in, live in," and the rescue reflected this attitude of "being one with the land, one with the people, . . . being the good shepherds." When questioned regarding the reasoning behind his company's involvement in the whale rescue, another participant stated simply, "We are part of the North Slope neighborhood." "People really do care about the world in which they live," and this world, of course, is bounded experientially rather than geographically: "All these kinds of things have some degree of personal effect on you and you can't say it's impersonal or you're not affected by it; you can't be halfway across the state and say, 'Oh my, it's too bad that those other folks had this problem.'" And this was equally true of those participants and observers from outside the state of Alaska: "The three stranded whales had the sympathy of the world." "There were numerous phone calls from all over the world, literally, internationally as well as nationally . . . offering all sorts of inventions and assistance and ideas," indicating the "un-distancing" by which the whales became part of our world despite our lack of proximity to the situation; and in this case the process of de-severance was aided by the television coverage since "you could [literally] sit in your living room and participate in the thing." A child's letter expressed the role of de-severance best of all: "Just because we are far away doesn't mean we don't care."

And concerned absorption was the appropriate human response. "Truly compassionate people don't think before responding in an emergency situation. They simply jump in and provide what help they can for as long as it is needed." In the case of the whale rescue, "a few people started it and once it got going nobody thought about anything; it just happened"; a forty-five minute conversation between industry and Borough officials "was all it took and from that point on there was no turning back," just as among the Inupiat there were "no lengthy discussions" and no consideration of how much money should be spent or how many days devoted to the effort. "They [all] just jumped into it like anybody would"; "it was just as simple as . . . they're trapped, let's go do something about it." "You just do what needs to be done." One participant suggested that even though ethical deliberation does not play a significant role in such concernful absorption, the outcome of such deliberation may well be the same response:

I don't think anybody sat down and analyzed, "Do I have an obligation to do this?" I think it was more, "Well, here are these three whales and we've got this equipment and let's help." . . . Later on when these guys were asked all these questions by the reporters, I'm sure they sat down and became more analytical and said to themselves, "Yeah, that was the right thing to do. Yes, we did have an obligation, it was the ethical thing to do." . . . I almost bet you they never really stopped to think about it. And then pretty soon everybody was caught up.

One participant noted of the critical reaction to the mounting costs of the effort that "there was nobody . . . who said, 'Wait a minute guys, let's put this down on paper and let's do some economic assessments of what we're getting ourselves involved in'; . . . we just said, 'Hey, let's do it.' I suppose if we had done that, maybe we wouldn't have gone up there." Acting out of concernful absorption, out of involvement with the whales as significant fellow beings in their world, participants "weren't looking at the bigger picture, [and] nothing else existed at the time; . . . getting them out . . . was all that matter[ed]."

Therefore, the rescue effort was the appropriate course of action because of its multifaceted conformity with our "ownmost and most proper" way of being. Perhaps the ultimate justification for the rescue effort, as well as its most significant aspect, is this evidence it offers of our being truly human; in the words of one participant, "I guess what it shows that is important is that mankind hasn't lost its humanity. When we get too callous to either man or animals, then it may not be a world any of us would like to live in."

What was the significance of their being *whales* rather than another type of animal?

"Whales . . . [are] a [ready] vehicle for that connection" we experience with other beings in our world, the connection that comes of our acknowledging that we are all fellow beings and that leads us to engage our best human instincts on their behalf.

(1) Unlike many other animals, whales "matter" in our society.
The fact that the three trapped whales were part of the "world" of significance for many participants and observers is due in large part to the influence of the public interpretation, which is becoming more environmentally sensitive in general and more concerned for whales in particular: "People are willing to spend more time and effort now to try to save animals," and

"the whale is the biggest mammal," not only in size but also, and more importantly, in popularity: "no group of animals has had a higher profile of the past several decades." "The worldwide attention that focused on the three grey whales also mirrored growing public interest in the fate of whales generally." Increasingly over recent decades "They" have become appreciative of and protective of whales. That "one" just doesn't kill them at will anymore is reflected in the widespread international opposition to commercial whaling and in the controversial nature of the aboriginal exemption from the moratorium; and this changing social norm influenced the public response to the trapped whales: "Just the fact that so much has been written about whales in the past twenty years: how they are very intelligent and how they communicate. And people go to great efforts, as tourists, to see whales out in the water. So that all contributes to the facts. So maybe they wanted to save these because of all they read about and think about whales. They're not just for oil anymore." While "a hundred years ago when you thought of a whale you probably thought they were light for your lamps, lubrication," mention of whales today brings to mind either wonder at their mystery and grandeur or outrage at the cruelty of the harpoon and the factory ship; 1988, in fact, was not only the year of this whale rescue but also "the first time in hundreds of years that no whales [were admittedly] being killed for commercial purposes." "There's a whale mania already in place, and you put a trapped whale in place, you get international hysteria."

"Whales occupy a place in the Western heart, even the hearts of the Eskimos who still hunt and revere them, that is special," and, in fact, local norms are probably even more appreciative of and interested in whales: "Whatever mystical response the huge animals trigger elsewhere, the feeling [in Barrow] toward whales is more intense and more intimate." The Inupiats' "whole lives are wrapped around the whales," and "there's a tremendous revere of the whale by the Eskimo . . . a deep reverence for that animal; even though these were gray whales it's still the same."

(2) Whales are uniquely able to invoke in us the wonder and awe that go hand-in-hand with respect.
Whales "are unmatched invokers of awe," and awe and wonder are the cornerstones of respect. "There's definitely this feeling of wonder" when you see "those beautiful animals . . . trapped out there, and you just can't walk away." One on-site observer suggested just this link between the experience of wonder and responsiveness:

> One of the first things we did was Search and Rescue took us out that evening to the hole. It was night and there was this really extraordinary scene: there was the ice, . . . the place where they were breathing, . . . this huge orange moon on the horizon. It was a very unearthly scene. You get there and there's this slushy pool and these three come up; and I remember really feeling . . . I could see why people were interested at the scene at that time because here's these animals, they're incredible animals anyway. . . . I don't think most people ever get tired of seeing whales. So here they are, very close, coming up.

For many of the participants and observers, then, "it was just such an incredible experience; it's almost indescribable" to be "that close to a species man doesn't get close to often." Whales are "amazing creatures," "magnificent creatures," and part of their awe-inspiring ability lies in their incomprehensible size: they were "big suckers, big . . . and we were just seeing a little bit of them," and "it was incredible to see these huge creatures, . . . to be that close to such a large mammal." "There's something otherworldly about them—you don't see them that much—and they're so graceful in the water." Most of us "love whales because of their amazing facts," and for some observers and participants this appreciation for their Being was enhanced by listening to their breathing and watching them spouting. In this light, time alone with the whales was "real special, the neatest portion of the whole thing." Such wondering appreciation also helped keep up commitment and energy levels on the ice:

> As soon as that first slab would go under . . . a whale would pop up in there and the Eskimos just laughed. It was really cool. . . . If the whales hadn't been popping up like that—I think that this was the key; the more they popped up the quicker the crews worked because they wanted this to happen again. And the whales always obliged. If the whales had just seemed lethargic . . . they probably wouldn't have been fought for as hard. . . . If they were just sort of half-sick-looking lethargic blobs back in the first hole . . . I don't think it would have worked. It was that playfulness on the whales part and the connection . . . that spurred it on and got that energy going.

(3) It is particularly obvious that whales are fellow beings and particularly problematic that we have seen them as resources.
The fact that these animals were whales doubtless made the caring response come easily: "even if you never saw a whale you would somehow feel connected to it." With whales, more so than with other animals, it is just so obvious that we are fellow beings and that viewing them as resources is highly problematic. If we have "some sort of soul mate in nature" it is certainly the

whale; like ourselves, they are intelligent and emotional and playful; they too form bonds of family and friendship, and they too have few predators and are thus largely unconstrained in their domain. Whales are "in so many ways so much like man," perhaps "even more intelligent than we are in their own world." Certainly in the case of the trapped whales, it was apparent to just about everyone that "there was an intelligence there," that "they wanted to be free . . . [and] knew we were their only hope," that "they were happy to have someone there who cared about them." Comforting them with a touch and a reassuring word and speculating that "they are mad be-cause they have to stay ther [sic] so long," participants and observers cer-tainly had no doubt that the whales, like themselves, were beings in and of and for themselves, not "things"; as one worker on the ice expressed it: "here is someone, . . . it's an animal who knows what's going on . . . and is aware of its difficult plight. And you just want to tell the guy, . . . 'keep alive for a while, you'll get out.' " Accordingly, while "the human race has been totally responsible for the elimination of wildlife in its care," nowhere has this consequence of the technological attitude been more apparent or more problematic than in the case of whales: "we have just decimated them," and

> news about the large brains and complex social relationships of some toothed whales, the remarkable songs of humpback and bowhead whales, and the awe-some diving capabilities of sperm whales has made a deep impression. Among the many examples of how we have greedily and mindlessly plundered nature's provender, none offends our sensibilities more than the destruction of the world's whale stocks.

What was the basis for opposition to the whale rescue?

(1) The public interpretation is, in general, Cartesian and anthropocentric, and in Alaska it emphasizes noninterference with natural processes.
Social norms strongly influence how we respond to other beings and to sit-uations like this, and in our society "They" "continue to assume that our problems and our welfare are somehow more important than any other creatures on this earth"; "there is a prevalent vanity in our way of think-ing that, because humans are the supposed supreme mammal, we are the most important living organism[s] on earth." Some of the participants and observers, influenced by the predominantly anthropocentric public inter-pretation, objected to the rescue effort on the grounds that "they're not of your species, they're not human" and that human needs must therefore

take precedence. "The idea that it was going to happen to three whales certainly didn't bother [them] as much as [a person] drowning [would]" because they "just don't equate whales with people." Similarly, the editorial cartoons criticizing the rescue—suggesting that the money would have been better spent on homeless and hungry people and that our concern should focus instead on, for example, hostages long "trapped" in foreign countries—grew out of the (Cartesian) social norms which encourage not absorbed involvement but detached deliberation. From this perspective, the would-be rescuers should have "thought first" about other potential uses for their time and money and, of course, should have concluded such deliberation with the judgment that human needs take precedence; they should have stopped and asked themselves, "Are you really going to go . . . up there and spend that much money . . . or is there something better you could do with that money?"

Similarly, local social norms rendered Alaskans "more comfortable with the idea that these things happen" and thus "lots of people [there] wondered 'What the hell was this crazy country thinking about?'" "There were far more phone calls from outside the state wondering what was happening than . . . from [Alaskans]." Not only would "most Alaskans . . . not think [whales] are equal [to people], [but] . . . the likelihood is that there would be more people than outside who would say, 'Let nature take its course.'" In Alaska, "They" say that "you should [not] go out of your way to intervene in the natural process," and "They" accept that "cruel death and dying are the natural state of things." In determining the most "fitting" response to the situation, then, many Alaskans thought that the noninterference aspect of "saving" should take precedence over the preservation aspect: Here's something in trouble . . . and you don't like to see something struggling or die, . . . but on the other hand . . . you don't rescue everything in the world." "People [there] who like whales would [say], 'That's really sad,' but [it is unlikely] that any of that would have happened if it had just stayed on a local or state level."

(2) Many people believed that respect dictated putting the whales out of their misery.
"Just like you would pull the plug on your father or your mother at some point" out of respect for their being human, in an effort to prevent their being consumed by dehumanizing pain and humiliation, some people saw putting the whales out of their misery as the more respectful alternative. Thus, some participants and observers suggested that "the most humane

solution may well be to shoot the whales"; in the words of one Barrow resident, "I suggest mercy killing. This is awful." From this perspective, preserving the whales' dignity in the face of "their obvious and growing misery" dictated not a prolonged rescue effort with questionable success but "kill[ing] the animal[s] and ending [their] suffering"; as one detractor expressed this position:

> My . . . reaction was, "Why doesn't somebody kill these critters and put them out of their misery?" My personal reaction was you should put them down like you put a race horse down [because] there's no hope here. I wish they could get away, but there's no way they're going to get away; even if you get them out of sight they're going to die in the ice somewhere. Let's just execute them and save them the grief.

Some of the rescue's primary supporters, in fact, also questioned whether the rescue effort or a quick death was actually more respectful of the whales: "It's very frustrating to see this and to have a feeling that maybe you're just prolonging their misery, maybe there is no way to save them, and maybe it wasn't the right thing to do."

(3) Many people thought that the motivations underlying the rescue were actually based on lack of respect for the whales.
And third, many of the detractors felt that the rescue was not in fact motivated by respect for the whales but rather was driven by our tendency to humanize animals and by concerns, such as public relations and the desire for a happy ending, which actually served to subordinate the whales themselves as means to these ends. In response to the naming of the whales, the media personification, and the not-infrequent comments to the effect that "It's like going out and freeing Bambi," detractors argued that "the whole point of whales is that they're big and wild and free, that in their own dominion they are as powerful as we are in ours. They are not big, fat people who swim well. They're whales. *To Bambi-ize or human-ize the noble beasts of the sea shows a lack of respect for what whales are all about.* Let the gray whales survive the ice, if they can, but let them be."

The fact that for so many people "it fits in with [their] idea of [nature] that there should be these whales, that if they get trapped people should save them, and that nature really isn't an ugly place," suggests a predominant lack of awareness of and respect for both nature and the whales themselves:

> [People] are completely alienated from [nature] on a day to day basis and their understanding of it is basically what they see in the movies. Walt Disney has

had more to do with the way the average American sees nature than that American's own personal experience. So naturally you get this kind of stuff, because what makes money is creatures that are basically just like you and me, so that's how Walt sold it and that's how people bought it.

"Most people only want to save Bambi-ized animals, like whales or baby harp seals. They're not interested in the sort of wildlife that isn't made into stuffed animals, much less in the hard, long-term task of preserving natural habitats," and in the minds of many of the detractors, this was the attitude that motivated the rescue effort, not genuine respect for the whales as wild animals.

And, similarly, detractors often noted what seemed to be evidence that the rescue was motivated by concerns other than the well-being of the whales. For example, even though "that just wasn't what the whales needed, . . . [photographers] really wanted to be on top of them taking pictures," suggesting that some members of the media valued the whales more as objects for a prize photograph than as beings in their own right. A critical editorial cartoon explained industry's involvement as motivated by public relations, and it was frequently suggested that shooting the whales to put them out of their misery was "quickly eliminated as an option" once the world became aware of the whales and the possibility of negative press arose: one observer commented, "I think once the video got out of this, there was a lot of 'We can't kill them now, it will look horrible.'" It was also thought that our desire for a happy ending rather than concern for the well-being of the whales themselves motivated the rescue, a position grounded in the premature declaration of success and in the decision against tagging the whales. From this perspective, the announcement that the whales had been freed on October 26 when in fact they would soon become trapped again in the refreezing icebreaker channel indicated that "the appearance of success was more important than actual success"; and the hasty departure from Barrow, without followup, was taken as further evidence of this position: "There weren't a lot of people hanging around to do stories about what these whales faced now that they were out of sight. It was like, 'OK, they're out of sight; we're free of this albatross; let's get the hell out of here.'" "People wanted it to have a happy ending, and so it didn't really matter how it really ended," thus the "collective fraud that we're all pretending it came out OK when no one really has a clue." And the decision against tagging the whales so that their progress could be tracked by biologists was also taken to indicate that the central concern driving the rescue effort was not respect for the whales, since "had those guys gone on and [had we known that they had] gotten stuck again, what was a remarkable effort really could have turned, have been soured." In the words of another detractor, "it sure wasn't about saving the whales or we would have done something to mark them so we could have found out if we really did save them."

What were the implications of this event for environmental issues in general?

(1) Respect for nonhumans and respect for other people are interdependent.
The unfolding of the whale rescue demonstrated that respect for the whales was closely linked with respect for people, in this case the Inupiat in particular. "If anyone was really responsible for saving the two whales it was the Eskimos: for just knowing the whales, knowing the ice, coming up with the idea of cutting the holes with chainsaws, which was basically the only one that really worked." Caring for the whales, then, depended on respect for the Being of the Inupiat, whose own Being, in turn, includes "respect[ing] each animal that we depend upon." Because "they've lived [with] and observed them for many years" and because "no one has more respect for our native animals than the local hunters," "we needed to listen to what they were saying and [follow] what they were doing" in order to succeed in the rescue attempt. In fact, several of the participants thought that the whales might have been freed earlier "if we would have just let the natives do their thing" instead of "shoving them out of the way . . . and assuming we had the answers: . . . the way white boys do everything." Respect for the Being of the Inupiat involved "work[ing] with their leadership" and acknowledging that "They know what they're doing. If we listen to them we will learn."

As one of several examples, when the whales ceased moving forward at the twenty-fifth hole in the series and workers "were out all night trying to force these guys to go," an Inupiat elder's attentiveness to the whales' behavior and his subsequent advice to "think like a whale" allowed rescuers to quickly ascertain and resolve the problem: "If you are going to help these whales, you've got to think like whales. Would a whale swim over a shoal so shallow that it thinks it is going to get stuck? No. A whale would swim around the shoal. That's what you've got to do, cut a trail around the shoal." "After two or three days of trying to coerce them to go, . . . as soon as a hole was cut in the right direction [in deeper water] they were in there," and "they were just getting so rambunctious and happy" "they just danced through these holes." In this way, respect for the whales was intimately linked with respect for the Inupiat people, a pattern with significant implications for our handling of environmental issues: respect for Being entails not only responsibility for the natural world and its nonhuman inhabitants but also appreciation for the local knowledge and perspectives that can often successfully be brought to bear on such questions.

(2) Much of the interest in such situations is at the level of mere curiosity rather than genuine concern.

The whale rescue also demonstrated how frequently "people don't really see what they are looking at," or how superficial curiosity rather than genuine concern may often be at the root of such interest in nature. The whales' plight "was just pure curiosity for some people," suggesting the tendency to encounter the world from the detached Cartesian perspective, without openness to wonder and astonishment. For many members of the media, "this was just another big event to go to," while for the public at large, "everybody wanted to go up there," in large part simply "because that was where it was at." And for several people the goal once they arrived in Barrow was to visit the whales, not out of wonder or desire for an intimate encounter but "trying to get a picture taken": in the words of one observer, "I'd seen the whales come up, so I ran over there . . . and I'm running from hole to hole and turning so [a colleague] can try to get a picture of me by the whales." An on-site photographer noted with some regret that his own encounter with the whales operated at this detached level:

> Ten days I was five feet from whales every day, and I never thought to touch them. You get into this mindset when you are working. . . . You aren't really part of it. You're always looking for the next picture. And I started dealing with *Time* and *Life* and they're wanting newer and better pictures. . . . It subdued somewhat what I might have done naturally if I was just there on my own to photograph this in a vacuum. I might have put the cameras down for a while and enjoyed it. But I worked pretty hard, and I never reached out and touched one.

In fact, it may well be that such curiosity drove the event from the beginning: "news is what's different," and "it was such an unusual thing: three whales trapped." One observer considered the source of our interest in the whales in this light:

> From my point of view, there is no answer that makes the American people or American journalism look very good. . . . We had at roughly the same time a couple of kids burn up in or near Barrow; it wasn't even a blip on the radar. I

suppose people get killed in fires in New York too but whales don't get trapped in New York Harbor; and that's why one's a national television story and the other one isn't. . . . Do I think that whoever the news editor was on the "NBC Nightly News" looked at that [original footage of the whales] and thought to themselves, "Gee, the people of the U.S. are going to be vitally interested in the fate of these whales"? I think he or she looked at it and said "Gee, hot video; let's put it on." And from there it sort of expanded. . . . A bunch of people say, "Cool. Whales. We'll go out and look at them." . . . That's sort of how this worked.

Whatever may or may not have motivated our interest in and involvement with the whales, it does seem to be the case that such events are characterized in part by curiosity, by the appeal of exotic situations, and that this approach may well drive out the possibility of intimate, wondering encounter. The environmental community would do well to consider the limiting factors at work in our superficial experience of the natural world and the possibilities for helping us all transcend these obstacles to more meaningful encounters.

(3) It is difficult to definitively interpret actions as reflecting either the attitude of "care" or the attitude of "technology."
Two particular aspects of the whale rescue demonstrate clearly the difficulty of determining with certainty whether a particular action truly embodies care or respect: the common desire to touch the whales and the decision not to radio tag them can both be interpreted in terms of "care" *and* in terms of "technology." At first glance, it would seem that not touching the whales suggests detachment and that the desire for physical contact indicates a sense of connection and openness to intimate knowledge. For some of the participants, it was indeed the case that the "best day of all" was when they first touched the whales and that they felt "moved . . . and scared for them, worried for them, concerned . . . [and] fortunate to be that close and involved . . . to be able to touch them." For others, however, touching the whales was something "you only do . . . once"—because it was there to be done—and not something that had any particular meaning for or effect on the individual. Some touched them "just for a picture": "They would come up almost like they were posing for a picture; they would sit there and you could put your hand on their nose so someone else could take a picture." And, interestingly, for many of the participants it was *not touching* the whales that explicitly embodied respect. One participant, for example, "would have liked to [have touched them] . . . but [didn't because she] believe[d] the less hu-

man contact the better for them," and for others connection and respect neither required nor allowed physical contact: "In their situation they were under the most stress in the holes and that's mostly when I saw people touching them. . . . I thought to myself they just don't need an additional source of stress. . . . I just had this personal experience of having eye contact with them, more mental connections with them."

Touching the whales and refusing to touch them variously revealed both respect and lack of respect, both genuine concern and mere curiosity, both connection and detachment; one participant expressed this very tension as it played out in her own response to the whales:

> I never did [touch them]. . . . In a way I just didn't want to. . . . Part of it was the sense that humans fondle everything. They weren't pets for Christ's sake! . . . You want to touch everything and you just mess it all up. . . . And part of it is you just grow up knowing you are not supposed to . . . touch, you don't get your smell on [them]. I say that on the one hand, and on the other hand I really did want to. You want to touch things; you want to know in every way you can what it is, what it's like.

And the decision against radio tagging the whales can also be variously interpreted in accordance with either "care" or "technology"; in the words of one participant,

> Sure they would have liked to have followed them, but I felt that people were wanting to do that more for the satisfaction of knowing for sure they made it; for me, I don't think that's the important part. Let's don't give them any more stress at the end. They were still in pretty bad shape when they were leaving; they'd gotten trapped in the ice again—in the slosh. . . . It was the last thing they needed.

Thus, tagging the whales when it was not in their best interests simply in order to satisfy ourselves of our success would have been using them rather than respecting them. But, on the other hand, for many "the appearance of success was more important than actual success," and thus the decision not to tag them could also have been driven, not by genuine concern for them but by our desire to believe that they survived regardless of their actual fate: the truth might have ruined the "story." "It was a fairy tale. . . . If you'd gone out and found them dead the next week somewhere . . . even the media would have tried to shut you up." The lesson again: it is necessary to look beyond appearances, beyond the surface of an action or choice, to the interpretation of Being underlying it in order to truly understand its appropriateness.

(4) Revealing as care is rapidly displaced by revealing as technology.
Even when care is the original motivation it can be displaced fairly easily by
the technological attitude; in this case, even where genuine concern was acknowledged as the original impetus it seemed that the rescue became less
and less about respect for the whales and more about other concerns. One
participant suggested the gradually altered nature of the commitment to the
rescue:

> Initially, the decision was made to help these . . . whales. You can't just let them
> die, watch them die, so you're going to do this thing. . . . By the time the impli-
> cation was realized, what that meant, it was almost like, "Well now that we've
> tried are we going to tell the world we failed and let them die anyway?" It kind
> of changed a little bit to where you felt an obligation to continue with the
> project.

Thus, an "effort that began as a humble attempt to help three stranded
whales . . . turned . . . and [took on] a life of its own": the spirit of the rescue "didn't seem to relate so much to the whales anymore as to the effort,"
and one participant noted, "I hope people realize that this thing ain't about
whales anymore."

"It had been started and it was like building the Panama Canal: 'We're going to do it. . . . We're the technological masters of the world and there ain't
nothing we can't solve.' . . . That really drove the thing. . . . They weren't sen-
timentally involved, they were technologically involved." In this light, rescue workers asked "when is that damn barge ever going to get over here so
we can go home"—not because the whales would then be allowed to return
to their interrupted migration; they were frustrated at the youngest whale's
death because "they had done everything they could to save [them] . . . and
then to lose one was kind of a failure"—not simply because of the death itself; and they "huddled around the pool, concentrating on efforts to install
de-icers in the water . . . almost oblivious to the whales as they bobbed gen-
tly, lit by the headlamps of snowmobiles."

And as the technological attitude becomes predominant, the experience
of wonder and awe recedes further and further beyond reach; in the words
of one observer, "It did move into the effort much more than the whales at
a certain point. . . . I felt really lucky that I had gotten there so early, that
there had been this very intimate [encounter], seeing them without all the
hoopla, because eventually it was just a huge circus . . . that did supersede
the whales eventually." "By the tenth day you didn't really care" and "after
two weeks it was really getting to be a big drag." Thus, "what seems odd

about this general reaction is not that it occurs from time to time but that it goes away so quickly"; in a world dominated by the technological attitude, care—genuine concern and respect for other beings—quickly gives way to the detachment that perceives other beings as objects whose well-being is easily subordinated to other concerns.

(5) The public interpretation gradually changes over time.
The whale rescue also demonstrates the dynamic by which the public interpretation gradually alters over time. The rescue both resulted from the public interpretation becoming more appreciative of whales and in turn contributed to its further evolution in this regard:

> Even in its small little way it had to do some good. . . . It brought people's attention to the plight of the whales. . . . They were already in view. . . . Had they not been, had [their endangerment] not been focused on and brought to the world opinion, this wouldn't have even been a story. But because they're viewed as an animal in danger, it became imperative to save these three. . . . It was kind of the finishing touch to a lot of hard work by people who were trying to save these animals, to get world opinion to say it's important to save these three.

That "a species . . . often associated with eating whales, killing them, [could find] some reasons within to reach out . . ." attests to the changing public interpretation regarding the relationship that should exist between human beings and whales. And, with respect to this event's own contribution to altering what "They" say, "One might forget that . . . people are driven as naturally to preserve as to destroy . . . were it not for the occasional endangered animals who raise their heads from the lower depths and cry for help." In this light, the rescue as a whole as well as such specific acts as speaking with the whales serve as marginal practices, often dismissed as old-fashioned or trivial, that remind us of "caring" ways of relating to other beings; the memorable photographs of people reaching out to touch a whale with expressions of awe and wonder on their faces capture the essence of Being, both our own and that of the whales, and thus help us achieve new ways of understanding ourselves and other beings; and serious reflection on the event's meaning, rather than "writ[ing] it off . . . [as] a dumb story," "can matter as much as anything, [can serve as] a trigger point . . . [or] touchstone . . . to lots of things that we're ignoring in our rush to improve our lives and have fancier toys." In each of these ways, then, this event, as "a time in which God's creatures knew a common bond," opened up new ways of understanding and responding to our world and thus contributed to the ongoing evolution of the public interpretation. It serves as a good example to the environmental community of the forces that help shape what "They" say regarding our place in the natural world.

(6) The intensity of the responsiveness in such situations suggests the alienation with nature as a whole.

This event also served to highlight, by way of contrast, the alienation that has largely come to characterize our relationships with nature and with each other. In this light, "understand[ing] why people feel connected this way [may shed some light on] what we've lost, what's there that's not being tapped into daily." In a way it is "pitiful that two whales who would die a natural death can take on such symbolism . . . [because it suggests that we are all] well beyond hearing and seeing what's around [us]." When we have "withdrawn into cities of concrete . . . and people [visiting Alaska] sit in their Winnebegos watching TV, . . . [when] getting away from it all [is] taking it all with [you]," when we are "alienated from [nature] on a day-to-day basis," it is "not surprising at all" that a single encounter with nature such as this should evoke such an intense response; "people are deeply conflicted about their relationship with nature, and it pops up at the weirdest times." And the widespread public response to the whales similarly highlights the extent of alienation within our own society:

> Here's a simple story that we can all respond to and we can all talk about, get excited and hopeful about. . . . We want to be connected to each other and to the world, . . . [but] what else do we share? . . . Feeling like we're personally caring about something that everyone else is caring about in some odd way is the way we generate community. Because it's so hard to do that anymore, we end up doing it through these national images. . . . It's sort of like a secular prayer or something in a way.

Weariness sets in for both whales and workers as the rescue effort drags on, and Bone, the youngest and weakest of the three, succumbs to illness and exhaustion and disappears beneath the ice. One of the whales props his abraded head on the edge of an airhole, resting.

Freedom in sight. The Soviet vessels break through the pressure ridge at the edge of the icesheet, creating an avenue to the open leads to the west as the whales continue to cross the ice through the series of airholes.

Bonnet and Crossbeak surface dangerously in a rapidly refreezing icebreaker channel. They are last seen diving in the direction of the open leads on the morning of October 28.

Although the whales' ultimate fate remains unknown, this encounter with Bone, Bonnet, and Crossbeak was, for many, a significant and unforgettable experience of human–nonhuman connection. In the words of President Ronald Reagan, "This dramatic rescue effort . . . reminded all of us of something essential about ourselves and our human nature."

Part IV Conclusion

I could never quite really figure out whether we all kind of collectively lost our minds or if we actually did something good.

Chapter 8   Recap and Reflection

In this chapter we will recap our study and consider some of the conclusions we can draw from it in light of the three objectives that were set forth in the Introduction. As we take a careful look at our results here, we will not be able to separate out those conclusions associated with each of the various objectives; the role of the case study cannot be separated from consideration of the influence of the philosophical traditions as conceptual lenses or comparison of the dominant and alternative frameworks because the latter are grounded in the former, and the influence of the traditions as lenses cannot be separated from the comparison of the frameworks because it is in the contrast between the frameworks that such influence is made apparent. Our purpose here is not only to recap the highlights of the study, condensing it to its very essence, but also to tease out and make explicit several of the themes that have emerged and questions that have been raised as we have progressed through our exploration of the three frameworks. Table 1 summarizes several elements of the lenses in contrast with one another.

Table 1. Defining Characteristics of Three Frameworks

	Lens I	Lens II	Lens III
essential human nature	rationality	reason + emotion responsiveness compassion	wonder awareness of Being "field of care"
essence of morality	resolving conflict among rights /interests what is just/fair?	care = + responding to need + avoiding harm what is help/harm?	care = respecting Being + noninterference + protection what "lets be"?
essence of agency	duty as moral agents principled logic	responsibility with respect to relationship attentiveness empathy	responsibility as "shepherds of Being" appreciation attentiveness
process of agency	detached deliberation characterize issue abstractly .	connected deliberation and response attend to context, detail, individuals experience	involved, connected action appreciate other's Being experience
relevant determinant	moral standing	particular relationship	what is "fitting" to our Being as care
relate to others	universal impartial	gradient of relatedness	"un-distancing" of significant others

Recap of Traditions as Lenses: Different Answers

The objective of exploring the influence of philosophical traditions as conceptual lenses was pursued primarily through the parallel recounting of the three versions of the whale rescue story. Reconstructed by an imaginary ethicist successively adopting the perspective of our three traditions, each of the three stories was unique and each brought to light both new aspects of the event and new ways of understanding previously considered aspects. To facilitate a final comparison of the traditions as conceptual lenses, let us briefly summarize the three versions of the event.

Whale Rescue Story I

The whale rescue was recounted in the context of extensionism. Given their inherent value, our dealings with the whales were understood to be subject to ethical rather than merely economic constraints, and moral principles were thus to be applied consistently regardless of species membership. The rescue effort was a matter of principle and universal obligation; our duty to help the whales was variously grounded in respect for their inherent value, in consideration of their suffering, in the compensation for past wrongs the species is owed, and in dismissal of the opposition as irrational. The whales were seen to have an abstract, symbolic value, and, additionally, were granted moral considerability on the grounds of their sharing with us the relevant criteria. The rescue effort was variously opposed because it was seen to be irrational (driven by sentiment rather than reason, out of proportion and offering no chance of success, and inconsistent with the rest of our actions toward whales, other nonhumans, and the natural world), because the conflict of interests it represented between human and nonhuman needs was thought to have been inappropriately resolved, and because it was seen to result from a misplacing of the value that should reside in the population or in natural systems. Seen in this light, the whale rescue, as a microcosm of human–environment interaction, suggested the common perception that human and environmental concerns are opposed rather than integrated, the prevalence of the anthropocentric mindset (even among supporters), and the questionable value of casting environmental appeals in sentimental rather than rational terms.

Both supporters and detractors relied on reason in their arguments, appealing to consistency and denying emotional involvement. Hierarchical dualisms were at work in the frequent "line-drawing" between whales and other animals (because whales were seen to be more humanlike and thus more deserving of moral considerability) or between people and whales (because the whales were not, in fact, people—the anthropocentric position); in the debate over individuals versus systems as the locus of value; and in the linkage between reason's valuing the system and emotion's valuing the individual. There was little attention to detail or to the particulars of the situation in this abstract recounting (for example, the fact that the whales had names was not even noted in this version of the story because the relevant fact was simply that they were whales, not that they were unique individuals) and no legitimate role for personal experience (we heard nothing but disdain here for the notion that touching the whales, speaking with them, looking in their eyes informed the judgment to act). Because the obligation to help the whales was understood to be universal, widespread support of the effort was to be expected and the participation of whalers was cast as a failure of reasoning (if whales are in they are all in). Anthropocentric objections to the rescue were dismissed as evidence of species partiality, as was consideration of shooting polar bears to save the whales. And both supporters and detractors were seen to cast the ethical debate as a conflict of interest (between humans and nonhumans, between individuals and systems) which demanded prioritizing competing claims.

Whale Rescue Story II

The whale rescue was recounted in the context of ecofeminism's care ethic. The rescue effort was not a matter of moral obligation or duty but of compassion and responsiveness in the face of need. Our responsibility to help the whales was grounded in our having been the cause of their oppression and in the various types of relationships involved: the subsistence relationship between whales and the Inupiat people, the immediate relationships between the whales and the participants and observers on the ice (founded in face-to-face encounter and fostered by daily interaction), and the media-mediated relationships between the whales and people across the country and around the world. The whales were thus not seen as distant objects of dispassionate ethical deliberation but rather were understood to be beings in need with whom we had relationships, relationships that in turn fostered a sense of responsibility for their welfare; empathy, emotional attachment, compassion, communication, and sympathy were thus all taken to be relevant and meaningful elements of the rescue effort. The significance of their being whales was explained in terms of the strong emotional attachments people readily form with whales (due in part to the combination of similarities to and differences from ourselves they embody) and the ease of experiencing empathy with them and identifying with their situation. Opposition to the rescue was variously grounded in the judgment that it constituted harm rather than help, in the appeal to the broader ecological context, and in the frustration with our failure to respond similarly when faced with human needs (especially given the presumption of a closer relationship with members of our own species). As a microcosm of the human–environment interaction, the whale rescue suggested both the possibility and difficulty of expanding localized concern to a more generalized level, the significance of particularity (as opposed to abstraction) in invoking responsiveness and compassion, the role of potential effectiveness in determining the range of the caring response, the motivational power of emotion, and the effectiveness of visual images in generating concern and motivating response.

Rather than an either-or approach to value in the natural world as posited by an expanded but still discontinuous moral community that maintains separation between those who are in and those who are out (on the basis of their similarity to the human model of considerability) and between moral agent and moral subject, the central element here was receptivity to the experience of connection and a corollary appreciation for the whales in all their similarities to and differences from ourselves. Reason and emotion were seen to be complementary human faculties and a central place was granted to empathy, sympathy, emotion, and personal experience as both significant motivators and legitimate grounds for ethical response. Relationships with the whales were understood to engender responsibility and compassionate responsiveness and to provide for potential effectiveness of response, and it was frequently detachment or the absence of relationship (the feeling that one could not communicate with the whales or empathize with them) that undergirded opposition; we also saw how relationships with these particular whales might be the source of a more generalized concern. Attention to the unique details of the situation replaced abstract characterization (we learned the whales' names, we saw Bone's playfulness and illness, we considered the role of photographs and of the Inupiat cultural/religious beliefs), and there were no appeals to principles or demands for consistency; similarly, the issue was not cast as a conflict of interest requiring some sort of ranking for its resolution. Responding to the whales' need and avoiding harm rather than dutiful respect for their rights were the central ethical concepts; and attentiveness to the particulars of the situation (to the whales' health, for example) was understood to be the vehicle for determining the appropriate response.

Whale Rescue Story III

The whale rescue was cast in the context of Heideggerian environmental phenomenology. The effort was explained in terms of the stewardship of life, which is our responsibility as human beings; such preservation of life was seen to be a "care instinct," a fundamental part of human nature. We recognized the whales as fellow beings (in and of and for themselves) and ourselves as "shepherds" rather than "masters"; we were aware of and responsive to impending non-Being; we promoted the unfolding of the Being of other beings; and we acted out of concernful absorption with the whales as significant members of our world. The whales were understood to be beneficiaries of the public interpretation (in accordance with which they "matter" in our society) and uniquely able to invoke the awe and wonder that accompanies respect for Being; and it was noted that as whales they facilitated the transition from the attitude of technology to the attitude of care (because it is particularly obvious that they are fellow beings and particularly problematic that they have been treated as resources). The rescue effort was opposed by people under the influence of either the Cartesian, anthropocentric public interpretation or the local norms that value noninterference with natural processes; detractors also argued that respect for their Being called for putting them out of their misery and that lack of respect rather than genuine concern for their well-being was the actual motivation for the rescue (for example, that they were being used as a means to human ends or humanized or both). As a microcosm, the rescue demonstrated the interdependence of respect for the Being of the whales and respect for the Being of the Inupiat, suggested that interest in such issues is partly at the level of curiosity rather than genuine concern, revealed the ambiguity inherent in interpreting actions in terms of either care or technology, demonstrated the displacement of care by technology and the process by which the public interpretation gradually changes over time, and highlighted the alienation that has come to characterize our "technological" existence.

Ethical guidance was thus seen to derive not from the process of deliberation, the moral standing of the whales, or consideration of principles and duties but rather from an understanding of the essence of our Being as care. Cartesian presuppositions regarding the role of reason and the superiority of human beings were replaced by the experience of wonder at the whales and of absorbed involvement with them as significant members of our world, and the public interpretation rather than the universal stance of the moral point of view shaped responses. Our being-in-the-world as a field of care shaped both our receptivity to the whales as "always already" related and significant and our "act first, think later" response; the whales were in our "space" whether by deseverance or in accordance with a more local sense of "neighborhood." Respect for the whales did not involve an assessment of their value as moral subjects but rather wondering attention to their Being. The event was seen to embody both the attitude of technology (curiosity, disrespectful "Bambi-izing," using them as means to a happy ending) and the attitude of care (acknowledgment of their Being in and of and for themselves, genuine concern and attentiveness, wondering appreciation), and debate centered on the two aspects of "letting be," questioning whether in this case care called for noninterference or for protection. And the effort was understood to result from a public interpretation that has recently become appreciative of and protective of whales and, in turn, to contribute to the ongoing evolution of social norms (by serving as a marginal practice and as a vehicle for artistic expression of Being and for reflection).

Different Interpretations

With the elements of each story now once again firmly in mind, and the differences between them now even more succinctly stated and apparent, let us take explicit note of the unique interpretation each lens offers of the following six aspects of the event; as an exercise to test familiarity with and ability to selectively apply the lenses, readers may wish to develop interpretations on their own before reading those offered here.

(1) "Certainly it is difficult to imagine a similar effort on behalf of other animals. Dolphins or porpoises or apes or koalas yes, but not menhaden or snail darters or timber wolves or warthogs."

Lens I. Whether or not we judge ourselves obligated to engage in efforts such as this rescue is determined by the moral standing of the nonhumans in question; whales, dolphins, porpoises, apes, and koalas, unlike menhaden and snail darters, are those species most obviously deserving of moral considerability given their joint possession, with us, of mammalian characteristics such as intelligence. The exclusion of timber wolves and warthogs, however, suggests a different and illegitimate sort of partiality at work in this judgment: although they too are mammals (sentient, intelligent, and so on) they lack the emotional appeal of the more charismatic koala bears and dolphins and are thus discriminated against and devalued.

Lens II. The essence of moral agency is responsiveness in the context of relationships, and so the possibility of relationship with the nonhuman in question serves as a crucial point of distinction: while relationships with cetaceans and other primates are not only possible but increasingly common, it is difficult to imagine a relationship with a snail darter or warthog. And part of the difference in responsiveness, no doubt, lies in our greater ability to empathize with nonhumans that are more like ourselves (apes and porpoises but not snail darters), in our related feeling that we can adopt their perspectives and thus know better how to respond appropriately, and in the emotional appeal of koala bears (versus the negative emotions often invoked by wolves, for example). Nevertheless, such a distinction does suggest problematic detachment and limited care.

Lens III. Although the essence of our Being as care lies in respect for other beings as the unique manifestations of Being they are, it is also the case that the public interpretation significantly influences what "matters" to us. As "They" have become more appreciative of the natural world at large in recent decades, beings such as whales, dolphins, porpoises, apes, and koala bears have especially captured our imaginations and thus have come to be recognized increasingly as significant beings in our world; "They" do not pay much attention to snail darters, warthogs, and menhaden (most of us, in fact, have no idea what the latter even is, so how could it have a place of significance in our world?), and "They" have a long history of active dislike for wolves. Additionally and relatedly, we know so much more about cetaceans and primates and we have a history of involvement with them: we have a stronger foundation for wondering appreciation and for awareness that we are all fellow beings; unlike koalas, for example, the beingness of a snail darter is much more difficult for us to grasp, in large part because of our relative lack of knowledge of them.

(2) The Eskimos cutting holes "took breaks to snack on bowhead whale meat stewing in a nearby warming shack."

Lens I. The participation of Inupiat whalers, like that of the Soviets, is ethically inconsistent and can only be accounted for as a hypocritical public relations ploy or as a failure of ethical deliberation. If whales are in the moral community they are all in, and ethical principles must be applied indiscriminately, without partiality toward individuals we happen to know or against those we happen to like to eat. From the nonanthropocentric perspective, whaling is extremely difficult to justify (the only counter-claim of any potential weight is cultural integrity, but this is generally not seen as sufficient to override the whales' claim to life), and thus the participation of whalers, who in essence view the animals as resources, cannot be understood as an acknowledgment of the moral standing of whales; and from the anthropocentric perspective only misguided emotion or the human interests at stake (such as public image) can explain the participation of whalers in the rescue.

Lens II. The fact that rescue workers were helping to save these whales while eating recently killed whales is not particularly difficult to understand. In fact, it is *because* of their subsistence relationship with the whales that the Inupiat people responded to these three; their intimate association with whales gave them a strong basis for empathizing with these three and rendered detached unresponsiveness particularly problematic. Treating the whales well, in fact, is seen as one aspect of ensuring the ongoing success of and balance in the subsistence relationship. And, of course, consistency in and of itself is not particularly significant; and any judgment regarding the unethical nature of the bowhead hunt (insofar as such a judgment is implied by seeing the Inupiat's role in the rescue as ironic) would have to take the cultural and nutritional context of Inupiat life into account.

Lens III. Our relating to the world through care often renders it appropriate that we protect other beings, but it does not imply that it is inappropriate for us to use the natural world to meet our needs. Respect for the Being of the whales is reflected both in Inupiat whaling and in this whale rescue: the Inupiat acknowledge that they and the bowhead whales they hunt are fellow beings together and the hunt is based on this attitude of care rather than on a sense of superiority and the right to exploit nonhumans as "things." It is appropriate to our being as care that animals who are needed be killed with respect and that those who are not needed be treated with the same respect and thus protected when necessary. Whales are significant beings in the world of the Inupiat, in large part *because* of their shared dependence on the Arctic environment.

(3) "We couldn't believe the national press coverage and the interest we got. It was probably the most incredible part about this whale rescue story: what it generated, the interest it generated worldwide. It [does] say something about us as human beings, doesn't it?"

Lens I. Media coverage of the rescue played a role in the widespread response to the whales by providing information as an input to the deliberation process; through the media coverage, people at a distance had access to the same information as did those at the site, thus contributing to the universality of the judgment as to the rescue's appropriateness. In contrast, the sentimental motivation experienced by many people who failed to deliberate properly was no doubt enhanced by the emotional tone of much of the reporting and by the pity-invoking visual images.

Acknowledging the whales' moral standing and acting in accordance with their right to life was the duty of all moral agents as such; respect for life is a universal principle, so it is to be expected that all agents, regardless of their involvement with the situation, their proximity to the whales, or their personal experiences would respond accordingly. What this widespread interest says about us as human beings is that our ethical judgment is universal in nature.

Lens II. Media coverage of the rescue played a role in the widespread response to the whales by providing a mechanism whereby people remote from the scene could nevertheless experience a personal relationship with the whales. Especially through television, people around the world were able to encounter the whales virtually face-to-face, almost as if at first hand, and it was this relationship that motivated their compassion and responsiveness. And, of course, visual images of the whales contributed to the emotional and empathetic aspects of the worldwide response.

Responsiveness to need, empathy, and compassion are at the heart of moral agency, and thus anyone who experienced a relationship with the whales would have responded accordingly. And such response easily spread beyond the immediate area because media coverage allowed people throughout the world to be involved with the whales, to witness their plight as if at first hand, and thus to experience a personal connection with them. What this widespread interest says about us as human beings is that we have a strong capacity for compassion and for the experience of connection (even in mediated relationships), especially when the other is a particular, identifiable individual on whom we can effectively focus our concern.

Lens III. Media coverage of the whale rescue played a role in the widespread response to the whales by enhancing the process of de-severance (bringing the whales directly into our world even where our Being as fields of care might have failed to encompass them as significant beings in our world). The media similarly made quite explicit the contrast between the true Being of a whale and the diminished Being of these three and thus engaged our instinct to act in such a way as to promote the unfolding of their Being: images of large freedom-loving whales confined in small holes and images of strong and vibrant creatures resting on the edge of the ice in weariness made the threat to their very Being clear even to the most inattentive observers.

Worldwide response to the plight of the whales was to be expected. Throughout the world, whales have increasingly come to be acknowledged as significant beings; they are admired, appreciated, and respected—"one" cares about their fate, as species and as individuals. Interest in these whales was thus the natural product of the public interpretation. And, of course, geographic proximity is irrelevant to our being as care: through the process of de-severance or undistancing, the whales were "brought near" us all. What this widespread interest says about us as human beings is that we are "fields of care," absorbedly involved in-the-world and concernfully attending to our fellow beings.

(4) "In fact, once the animals were given their nicknames, the option of shooting the whales was replaced by the possibility that the defenders patrolling the breathing holes might shoot anything that threatened them, including the polar bears lurking nearby."

Lens I. In no way would shooting the threatening polar bears be a justifiable course of action: like the whales, polar bears are an endangered species (if anything the status of the gray whale population is much more secure); if the whales are members of the moral community then so are polar bears, and although the whales may have a claim upon us for our assistance, they have no claim against their own predators and we have no duty to protect them with that degree of interference (unless, of course, a priority principle is posited in accordance with which the whales' interests override those of the polar bears *and* override the general principle of our not interfering with interactions among other species). And certainly from the perspective of ecocentrism, there can be no justification for saving individual animals at the cost of disrupting predation cycles. As this commentator perceives, to consider such a course of action would mean that emotional attachment to the whales was inappropriately swaying rational judgment.

Lens II. Although it is not a course of action to be undertaken lightly, it may well be that shooting the polar bears in order to preserve the lives of the whales, if it actually came to an immediate and otherwise irresolvable threat, might have been the appropriate course of action. It was the whales, after all, not the polar bears, with whom we had an established relationship: we knew the whales personally, whereas the polar bear threat remained an abstraction, and when a choice must be made the closer relationship justifiably has greater ethical "pull," without of course justifying detachment from the polar bears in making the decision. Attention to the broader ecological context, however, renders such action problematic; we need to be sensitive to the network of relationships between polar bears and whales in the context of Arctic predation and to respect how both animals fit into the larger ecological community.

Lens III. The central problem in determining the "fitting" course of action involves choosing between or balancing the two aspects of "letting be," noninterference and protection, and this judgment can only be rendered against the background of respect for Being. It may be that noninterference is the most caring course of action when protection hinges on lack of respect for the Being of other beings: in this case, then, it is appropriate to save the whales from the ice out of respect for their Being, but it is inappropriate to save them from polar bears (by killing the bears) because such an action reflects lack of respect for the Being of the bears. Although protection in the form of saving them from the ice lets the whales be whales without any corollary disrespect for other beings, protection in the form of saving them from polar bears lets the whales be whales but only at the cost of not letting the polar bears be polar bears. Thus it is appropriate to protect the whales, but only up to the point at which doing so is disrespectful of the Being of other beings (unless Nature itself has a Being that we fail to respect when we interfere with the processes of life and death in the natural world).

(5) *"[People] would come back and talk about touching the whales and it truly affected them."*
Lens I. Touching the whales, of course, had no place in the process of ethical deliberation; it produced no information relevant to the decision to attempt the rescue, and it was doubtless motivated by emotion and sentiment.

Lens II. Physical contact was one of the most significant elements of the relationship between the whales and the people on the ice with them. It served as a form of communication and thus went along with words of reassurance. For many people it fostered a greater sense of empathy with the whales. And for still others it was touching the whales (being so close to them) that gave rise to the concern and connection that motivated their ongoing involvement. Touching the whales was an important part of the relationship between the whales and the people on the ice, and as this relationship formed and grew, people were obviously changed by it.

Lens III. Touching the whales is a prime example of how a single act can reflect either "care" or "technology." Touching the whales with an attitude of wonder or a desire to know them intimately is part of our Being as care, and the experience undoubtedly left people with a deeper appreciation for the unique Being of the whales, a sense of awe, and a tendency toward wondering reflection on their beingness and on Being in general. Interestingly, though, the same attitude of respect could just as well lead to the judgment that touching was inappropriate. People who touched the whales "because they were there" or simply in order to have an unusual photograph or an exotic experience embodied instead the attitude of technology and, closing themselves to the experience of wonder, were not likely to come away from the experience changed in any way.

(6) *"They are the breakthrough species in the bond between man and animal."*
Lens I. The bond between man and animal is a recent extension of the bond between man and man: joint membership in the moral community. Whales

are in the vanguard of this extension because they so obviously share with us the criterion for moral considerability, be it rationality, sentience, self-aware experience, etc. They are like us, so it makes sense that they are among the first to be acknowledged as sharing moral standing with us.

Lens II. Unlike the so-called bond based on extension of the moral community—a bond that, in fact, is grounded in detachment on many levels—a true connection between humans and nonhumans is possible, especially with animals such as whales. Personal relationships between particular people and particular nonhumans help to define each of us as the individuals we are, just as interhuman relationships do. As the Cartesian worldview and its emphasis on the fundamental separation of self and other, human and non-human, culture and nature is displaced by a worldview that acknowledges interdependence and relatedness, those animals like whales with whom we can most readily empathize and communicate are the first we reach out to in search of relationship. And, interestingly, it is these same animals who, in their unique combination of similarities to and differences from ourselves, have contributed to the dismantling of the Cartesian worldview by posing an implicit challenge to the dualistic and anthropocentric assumptions on which it rests.

Lens III. Widespread experience of animals as significant members of our world, with whom we have a bond in that we are "always already" related to and absorbedly involved with them, depends in large part on the gradual alteration of the public interpretation away from its "technological" view of animals as resources and of human beings as "masters" of nature. Because whales are so obviously fellow beings rather than "things" and because they are uniquely able to invoke a sense of awe and wonder, they are among the first nonhumans that "They" acknowledge as "mattering." Similarly, "we look into the danger and see the growth of the saving power," and nowhere has the danger inherent in the attitude of technology been more blatantly apparent than in our "revealing" of whales as resources for us, given the decimation of their populations; having witnessed our own destructiveness toward whales, we are especially conscious in their case of the danger of that less than fully human mode of relating to other animals.

Different Questions

That each tradition-as-lens offers a unique interpretation of the same aspects of the event has just been demonstrated. Similarly, the structure of Chapters 3, 5, and 7 demonstrated that each tradition-as-lens offers unique answers when presented with the same general questions. Although necessary for the purpose of exploring the powerful influence of conceptual lenses, the parallel structure of the three "stories" doubtless obscured another equally important aspect of conceptual lenses: not only do they offer different answers to the same questions ("unpacking" the questions differently and holding different expectations as to what constitutes an answer, explanation, or justification), but they also pose their own unique questions and thus direct our attention toward different aspects of the situation as the relevant or meaningful ones. As the next step in our comparison of these traditions, then, let us explore this additional function of conceptual lenses: what are some of the specific questions each leads us to ask in the attempt to make sense of events such as this whale rescue?

Lens I leads us to ask:

1. To what extent are we/should we be willing to apply the same standards and rules to human–human interaction as to human-nonhuman interaction ("without prejudice")?
2. Are the parties in question (for example, the whales) members of the moral community? Do they have moral standing? Why or why not?
3. What is our obligation when faced with (animal) suffering that we have the ability to remedy?
4. What failures in ethical deliberation contribute to the moral disagreement over this issue?[1]
5. What is the locus of value: individuals or the larger systems they constitute (populations, ecosystems)? If the latter, then what is the status of the population/system in question?
6. How should we resolve conflicts of interest between the various members of the moral community? How do we justly distribute limited potential for interest fulfillment?
7. How do general principles apply to this particular situation? What principles (or other good "reasons") justify this course of action? What are the alternative courses of action (uses of the time, money, and so on), how are they justified, and what justifies this choice over these others?
8. Is this course of action motivated by sentiment or by reason? Is it the product of illegitimate bias?
9. What is the cause of/who is to blame for this situation? With whom does the obligation to address it lie? Given historical injustice, how do we fulfill our duty to make restitution?

10.What are the limits of ethical obligation? Short of a successful effort, at what point have we fulfilled our duty/are we justified in calling a halt to the effort?

Lens II leads us to ask:

1. What is our responsibility in this situation, given our history of oppression?
2. What is the nature of the particular relationships at work here, and what responsibilities accompany them? To what extent are the relationships ongoing?
3. In what way does "being there" affect moral judgment? What is the impact of face-to-face encounters with the subjects of concern? What is the impact of touching? Does physical proximity make a difference in the moral judgment/response? If not, what is the mechanism by which agents at a distance experience themselves in relationship?
4. To what extent does empathy factor in the decision-making process/in the response/in the level of the agent's commitment? Is this a two-way relationship? What type of communication, if any, occurs and how does it influence the response/commitment?
5. How does personal experience shape the response/decision-making process? How are participants changed by their encounter?
6. What about this situation appeals to emotion, and to what extent is this appeal a significant motivation? How do reason and emotion work together to determine the appropriate response?
7. What accounts for differential responsiveness from one situation to the next? What factors inhibit a sense of relationship/promote detachment? Do closer relationships imply greater responsibility/have more ethical "pull"?
8. What are the prospects for concern transcending the particular situation/set of relationships in question? What are the obstacles to such expansion of concern? How can abstract issues be personalized for increased responsiveness?
9. Is there any potential for effecting change here? Can the agent make a difference in this situation? To what extent is apparent apathy actually evidence of the feeling of powerlessness rather than an indication of a lack of the disposition to care?
10. What are the details that uniquely define this situation, and how do they affect the assessment of the appropriate response? What is the larger

context within which this situation arises, and how does it affect the assessment of the appropriate response?

Lens III leads us to ask:

1. Given our responsibility as "shepherds of Being," what is the "fitting" response in this situation?
2. What is the essence of the beings in question, and how is it being threatened by or expressed in the current situation?
3. What is the appropriate choice between or balancing of the two aspects of "letting be" in this situation?
4. Are the beings in question members of our world? If not, why not?
5. How can we learn more about these beings? (This is asked in the spirit of wonder, not out of curiosity or the desire to exploit.)
6. What do "They" say about the situation and about the beings in question? How has the public interpretation of this issue/these beings changed over time? What has contributed to this change? To what extent is the public interpretation still Cartesian in nature, and how does this influence our Being with respect to the case at hand? Do local norms suggest different responses/interpretations?
7. Does this particular encounter embody wonder, respect, and genuine concern or mere curiosity and lack of respect? Does the attitude of technology or the attitude of care motivate this response?
8. How does the event in question contribute to the gradual alteration of the public interpretation?
9. How do the individuals involved experience the encounter? How does previous personal experience shape their response to the current situation?
10. By what process does the attitude of care emerge given the technology orientation of the public interpretation? By what process does the attitude of technology come to displace the attitude of care as events unfold?

We should also note that these lists of representative questions are equally revealing in what they do *not* include. Certain questions are not asked because the answers are assumed: Lens I does not ask how people at a distance from the site came to judge the rescue as appropriate, Lens II does not ask whether partiality played a role in responsiveness, and Lens III does not ask whether the process of de-severance occurred. Similarly, others are not asked because they are taken to be irrelevant: Lens I does not ask about the influence of personal experience on decision-making, and Lenses II and III do not ask about moral standing.

But the Lenses Seem Similar—Are They Really Unique?

It is also fair to ask at this point, however, whether our lenses are actually just expressing the same concepts in different terms, whether the questions they pose and the answers they offer are actually the same substance in disguise, as it were. For example, we might consider whether the "duty" to balance the scales with respect to our historically unjust treatment of whales (Lens I) and the "responsibility" that comes with our having been the cause of their oppression (Lens II) are actually just different words for the same phenomenon: we caused it, so we have to fix it (or something to that effect). And, indeed, if our three traditions differ only insofar as they have different labels for the same concepts, then they are quite likely redundant and their function as conceptual lenses may be empty.

There is an important distinction to be made here, however: that two observers comment on the same aspect of the world does not imply that it means the same thing to both of them, and it is just such a difference in underlying meaning that gives each of our traditions its uniqueness and its power as a lens. To return to our earlier phenomenological example, where you see a seashell I may see an ashtray, and while we are both commenting on the same physical "thing," your calling it a shell and my calling it an ashtray is much more than our having different labels for it—it is, as a matter of fact, a different thing for each of us and the terms we use for it crucially embody that difference.

If the background chapters that sketched out each of our three lenses served any purpose at all, surely it was to demonstrate the fundamentally different starting points and assumptions of each tradition, and it is in this light that we see the difference between our "duty" to balance the scales with respect to our historically unjust treatment of whales and the "responsibility" that comes with our having been the cause of their oppression. Positing a "duty" or obligation on our part is the product of (among many other factors) an understanding of the human self as a moral agent engaged in an explicit deliberation process and adopting the moral point of view, a commitment to justice or to balancing the benefits and burdens among members of the moral community, justifications for expanding the moral community to include whales, determination of the nature and content of their "rights," and systematic prioritization of rights claims to resolve the conflicts of interest among members of this expanded moral community. Positing a "responsibility," in contrast, is the product of (among many other factors) an understanding of the human self as essentially embedded in defining relationships, which in turn impose responsibility for the well-being of all parties to the relationship; an understanding of agency in terms of responsiveness rather than detached deliberation; recognition that causal chains of oppression constitute relationships, even with distant or abstract others; and receptiveness to expanding the potentially limited realm of care beyond immediate relationships. Moral standing and a commitment to the principle of justice are the source of "duty," whereas the nature of the relationship is the source of "responsibility." Although in both cases our historic mistreatment of whales has implications for our present ethical behavior, the similarity ends there; "duty" and "responsibility" are not simply different labels for the same concept but rather are fundamentally different phenomena, uniquely and meaningfully defined by our conceptual lenses.

There is one other basis for questioning the uniqueness of our three traditions and the content of their function as conceptual lenses, however. The flip side (so to speak) of simply applying different labels to the same concept is invoking the same terms or images: if both Lenses I and II find the parallel with Jessica McClure meaningful and if both Lenses I and Lens III rely on the analogy between "pulling the plug on your father or mother" and putting the whales out of their misery, then we might ask once again if, in fact, the philosophical traditions actually function as unique conceptual lenses. Here again, however, we need to look beneath the apparent identity and recognize the underlying meaning. Lens I finds the parallel between the whales' plight and that of Jessica McClure relevant because it captures the very essence of extensionism: we judged it our duty to rescue the whales just as we had earlier judged it our duty to rescue the child (on the basis of their inherent value, their right to life, their suffering, and so on) without the partiality for our own species and the assumptions of human superiority that constitute anthropocentrism. It was explicitly noted in Story II, however, that "the analogy is apt only because of the feelings involved"; Lens II does not invoke the parallel because of the underlying identity of principles but rather because both events demonstrated emotional motivation and compassionate responsiveness to particular individuals in need. The meaning of the analogy is not at all the same for the two perspectives, and they neither interpret it nor utilize it in the same way.

And the same is true of the analogy to terminating the life support of one's parents. Whereas our first imaginary ethicist invoked this image to confirm

Figure 1. Positioning of Lenses along Continua

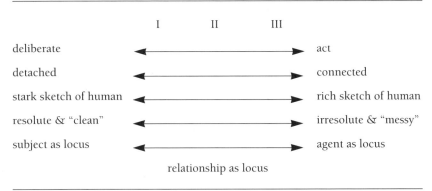

	I	II	III	
deliberate				act
detached				connected
stark sketch of human				rich sketch of human
resolute & "clean"				irresolute & "messy"
subject as locus				agent as locus

relationship as locus

the extension of moral considerability beyond the human realm, the ethicist of Lens III used it as an example of respect for the Being of other beings; "pulling the plug" is an appropriate response to prolonged pain and irreversible illness because it preserves human dignity or prevents dehumanization, and shooting the whales similarly would have been the appropriate response to their misery as a means of preserving their dignity or Being and preventing "dewhalization" (their being reduced to something less than what they truly were). Again, the purpose and meaning of the analogy are not at all the same in the two accounts, and the lenses are thus not more similar than we have come to believe. In fact, the uniqueness of each tradition's interpretation of the same element seems to demonstrate once again not the similarity of the traditions but rather their powerful influence as unique lenses.

Lenses Fall in Unique Positions along Continua

Having clarified the essentially superficial nature of the seemingly similar aspects of our three "stories," let us now take a brief look at some of the subtle points of contrast among the lenses. Given our second objective, exploring the adequacy of the dominant framework in environmental ethics and the potential of the two alternatives as supplements, we will be especially concerned here with the points of divergence between Lens I and Lenses II and III collectively and with the limits of Lens I and the contributions made by Lenses II and III (separately and collectively). Although many other contrasts could doubtless be uncovered and evaluated (some, perhaps, more fruitfully), here we will consider five that seem particularly relevant to our three objectives; before reading those offered here, readers may wish to develop their own explanations or to identify and explain additional points of contrast. Figure 1 graphically depicts these five continua and the relative positions of the three lenses along each of them.

Continuum of Deliberation/Action

Our three lenses occupy distinct positions along a continuum defining moral agency in terms of deliberation. In Lens I deliberation is the very essence of moral agency: human beings are moral agents because we choose, because we reason through the application of principles and the evaluation of consequences, because we think and are aware and thus can be held accountable. Accordingly, in Story I the imaginary ethicist focused our attention on the decision to attempt the rescue and considered his task complete when he had thoroughly probed the rationality (or otherwise) of the choice.

Stories II and III, however, accounted for the initial decision but also took into account other aspects of the event (naming the whales, touching and communicating with the whales, enlarging the holes into channels), suggesting that more than a single decision was at work and that agency encompasses more than simply deliberation. In Lens II deliberation is part of moral agency, but responsiveness and attentive presence are its very essence; that the agent will respond to need is virtually a given, and deliberation serves more to assess potential responses in light of the sense of relatedness. Agency is thus a blend of deliberation and action.

And in Lens III deliberation has little place in truly human agency: the Cartesian subject (whose Being as care is deficient) deliberates, but *Dasein* "acts first and thinks later." Human beings are agents because our being is in-the-world (meaning that we act concernfully) and because we are uniquely aware of Being; awareness here is not a prelude to deliberation but rather to appreciation, wonder, and absorbed, concernful involvement. As we move from Lens I to Lens II to Lens III, then, deliberation is decreasingly relevant and, correspondingly, our focus is increasingly broadened beyond the original decision or initiating action and our understanding of agency is increasingly action-oriented.

Continuum of Detachment/Connection

Closely linked with the changing understanding of agency (deliberation to action) as we progress from Lens I to Lens II and Lens III is the conceptual proximity between moral agent and moral subject. In Lens I agent and subject are fundamentally at a distance, even when both are acknowledged as members of the moral community; the human ideal lies in autonomy, and the moral point of view enjoins detachment in its requirement of objectivity, impartiality, abstraction, and universality. In Lens II agents and subjects are fundamentally in relationship, a network of relationships, in fact, that defines the agent; the agent does not judge from a distance but rather responds from the position of embeddedness within these defining relationships, and detachment—even where immediate relationships do not seem to be operative—is problematic. In Lens III the human agent is fundamentally in-the-world, "always already" related to others in a way that denies objectification and detachment (unless, of course, one experiences oneself as the less than fully human Cartesian subject in a world of objects) and in a way that is not limited by involvement in particular relationships (as "connection" in Lens II may be). And it should be noted that the "connection" of Lenses II and III is not the loss of self in merger with the other reminiscent of deep ecology; self and other are distinct, but they are fundamentally connected rather than detached and this connection is the ground and heart of ethical response.

The imaginary ethicist of Story I thus had no reason to acknowledge or explore any of the relationships between the whales and participants or observers, whereas Story II would hardly exist without them; likewise, it was the essential connectedness with the whales as significant members of our world, and the contrast between this "care" stance and that of the attitude of "technology" (wonder versus curiosity, beings versus things) that gave Story III its coherence.

Descriptive Continuum: Agent as Human Being

Taken together, the above two items suggest yet a third continuum: the increasingly rich characterization of the agent as human being offered by Lenses II and III. The agent as human being is rather starkly sketched in Lens I. The human ideal is rational, autonomous deliberation, and the agent either fulfills that ideal or he doesn't. The agent is basically (or might as well be) a disembodied will; universality, in fact, virtually assures the fundamental interchangeability of agents (in other words, that it is *me* rather than *you* making the decision is irrelevant to the process of deliberation).

Lens II, in contrast, offers a more detailed and richer sketch of the agent as human being. She is a particular individual with her own set of highly relevant personal experiences, emotional attachments, and needs. She is not only rational but also empathetic, emotional, and compassionate. She is defined by a network of personal relationships (with other people, with animals, with places, etc.) that influence the nature and content of her ethical response; and she is pulled in multiple directions by her desire to respond in the face of need and thus has no black-and-white ideal to guide her personal growth. The caring response for her and the caring response for me may well be different, even given very similar situations, in large part because of our individuality as agents.

And in Lens III the sketch of the agent as human being is richer still, although perhaps less individualized than in Lens II. Neither independent, disembodied will nor participant in a variety of particular relationships, *Dasein* is "always already" in-the-world. He is experientially immersed in multiple "worlds"—his own individual world, his professional world, the world of his community, and the world of his society at large, for example—each of which has its own (to some extent) unique interpretation of reality and set of norms and expectations. He is open to new interpretations of reality, new ways of understanding the world and his place in it; and he approaches the world and the beings composing it with wonder, appreciation, and reflection. His Being, then, is the product, not simply of particular relationships, but rather of multiple and interacting worlds and multiple and ever-changing public interpretations, as well as his own experiences. He, too, has no hard-and-fast ideal against which to measure himself but rather is shaped by and in turn helps to shape the various public interpretations governing his world.

And, of course, our imaginary ethicists drew on these very sketches of the agent as human being in recounting their respective versions of the whale rescue; this will become clear if we construct in our imaginations pictures of the decision-makers depicted in each story. We know (and need to know) very little about the agent in Story I who speaks of respect for life, the inconsistency of opposition to interference with nature, and the symbolic value of whales. On the other hand, in Story II we know (and need to know) about the agent's relationships and emotional attachments, her experience of empathy or communication with the whales, and her previous experiences in the natural world. And, similarly, in Story III we know (and need to know) about the norms that govern the agent's particular world(s), the experiences that shape his recognition of significance in his world, and the underlying attitude with which he approaches the whales and the other participants.

Continuum of Resoluteness

As was suggested in the previous item, ethical "answers" become increasingly irresolute as we progress from Lens I to Lens II to Lens III. Although Lens I does not always presume absolute standards, it does offer minimally "messy" procedures for ethical decision-making. It posits an explicit stance for the moral agent to take and an accompanying procedure for deliberation, both of which are embodied in the moral point of view; and it reduces dilemmas to the form of "A" versus "B" and then applies some sort of priority principle to rank the competing rights or interests. Of course, such prioritizing is not simple; the point is that it is thought to be possible. From this perspective, resolving ethical dilemmas is a matter of applying rules. So, for example, the whale rescue was cast as a conflict between human and nonhuman needs and was judged inappropriate on the grounds that human needs should receive priority—a controversial position, but nevertheless a resolute, "clean" one.

Lens II, however, offers no such "neat," virtually mathematical, mechanism for the resolution of ethical dilemmas. Principles cannot be applied universally given the significance of the details that render each situation unique; rather, the larger context as well as these details factor into the particular decision at hand, as does the nature of the relationships that give rise to the dilemma or that are affected by its resolution. Similarly, responses are to be tailored to the individuals involved, requiring attentiveness and inherently implying uncertainty, because an action on my part that would be a "help" to you may or may not constitute a "harm" to someone else; and the agent is "pulled" in different directions by the emotions and responsibilities associated with her various relationships and by her own needs as well. Rather than offering a fairly black-and-white answer justified in terms of universal principles, Lens II often must reply "it depends" when confronted with an ethical dilemma: it depends on the context, on the details of the situation, on the needs of the individuals involved, and on the nature of the relationships in question. Thus, the appropriateness of the whale rescue depends on details such as the Eskimos' need for the whales as food, on the context of a subsistence culture and a history of oppression, on the whales' ability to endure throughout the course of a prolonged rescue, and on the nature of the personal relationships between the workers and the whales, for example.

And "it depends" is equally ambiguous in Lens III. Determining the appropriate course of action proceeds through consideration of the appropriate mode of being, and that in turn hinges on its conformity to our Being as care. An action is wrong neither because it contradicts universal principles nor because it violates relationship-derived responsibility, but rather it is inappropriate insofar as the underlying attitude is one of technology rather than of care. Thus, not only is it extremely difficult to judge the appropriateness of another's actions (given the uncertainty inherent in assessing his attitude toward Being), but our only guidance lies in our understanding of our own Being as well as that of the other beings in question. Resoluteness thus depends on knowing ourselves intimately and honestly, knowing the Being of other beings with enough depth and appreciation to be able to distinguish acts that promote their Being from acts that disrespect it, and determining the "fitting" balance between or choice among the two aspects of "letting be" (noninterference and protection) as suggested by each situation. In this light, the appropriateness of the whale rescue depends, for example, on whether saving the whales or putting them out of their misery would best "let the whales be whales," on whether the effort used the whales as means to human ends (such as good public relations) or was motivated by a desire to promote the unfolding of their Being in and of and for themselves, and on whether care in this situation called for noninterference or active protection.

Continuum of Morally Relevant Determining Factor

And as we move from Lens I to Lens II to Lens III the reference point for judging the appropriateness of an action becomes ever more closely associated with the agent himself rather than with the subject. In Lens I, whether the subject possesses the morally relevant criterion (sentience, rationality, life, and so on) is of primary significance in the determination of the agent's duty: if the subject is in the moral community, status which is completely independent of the agent, then certain obligations are imposed on the agent and the appropriateness of his actions is judged in this light. The locus of the determining factor is thus the subject. In Lens II, however, it is not the nature of the subject per se but rather the nature of the relationship between agent and subject that largely determines the agent's responsibility; thus, the locus of the determining factor is the interpersonal, the relationship, and is no longer totally independent of the agent. And in Lens III, of course, the attitude or Being of the agent, whether he "reveals" another through "care" as the unique being he is in and of and for himself, determines the appropriateness of a given act, largely independent of the subject.

Thus, in the case of the whale rescue as recounted through Lens I, the principles at stake and our duty in light of them were a function of the moral standing of the whales, and they were variously thought to have such standing or not, depending on the anthropocentrism or nonanthropocentrism of the agent in question. Story II gave no role to the moral standing of the whales but rather argued the appropriateness of the rescue effort in terms of our relationship-derived responsibility for their well-being; different agents could thus be taken to have justifiably different levels of responsiveness depending on the presence or absence of a personal relationship with the whales and on the nature of the responsibilities implied by the particular type of relationship. And Story III presented the effort's appropriateness or inappropriateness on the grounds that it either did or did not comply with our Being as care, that it did or did not grow out of and reflect respect for the Being of the whales.

Although the fact that the trapped animals were whales did figure prominently in each of the three accounts (perhaps suggesting at first glance that Lenses II and III retain the "criterion" approach, which posits the subject as the locus of the morally relevant determining factor), the interpretation of their significance as whales in fact confirms rather than contradicts the changing locus. That they were whales figured in Story II in terms of the consequent ease of forming empathetic, two-way relationships and in Story III in terms of the wonder and awe we so easily experience in their presence; of course, their being whales also meant that the public interpretation was already operating in their favor (again suggestive of the "criterion" approach of Lens I), but this, too, is not a result of their possessing the traits associated with inherent value but rather of our having come to "reveal" them through care as significant beings in and of and for themselves.

Another interesting point is suggested by the various positions our lenses assume along this particular continuum (the subjects ± the relationship ± our Being): in this light, Lens III can be seen to resolve the often-noted tension in Lenses I and II regarding the respective limitations to universal and particular concern (recall that we discussed in Chapter 4 the inability of Lens I to account for "special ties" and the similar difficulty Lens II has in accounting for generalized concern beyond the realm of the given network of relationships). By locating the morally relevant determinant in the Being of the agent himself rather than in something external to himself (either the subject or the relationship) and by conceptualizing the human being as a field of care, Lens III virtually does away with both the problematic limitations in Lenses I and II and the tension between them on this issue. In his Being as care, *Dasein* may well make no distinction between the universal and the particular, between the remote and the close-at-hand; although clearly "one" has special responsibilities to one's family and friends (only the occasional philosopher has ever questioned what "They" say on this point), there is no corollary limitation on care because the agent is also a field of care, relating to distant others as care through the process of de-severance.

Recap: Dominant versus Alternative Frameworks

Finally, the underlying impetus driving the exploration of alternatives in light of their similarities and differences is a judgment regarding the inadequacy or insufficiency of the dominant framework on its own. And because one of our original objectives involved just this assessment of the need for and possibilities of supplementing the dominant tradition, let us now turn our attention to an explicit recap of this question as our case study has shed light on it (of course, the burden of the argument in favor of supplementing the dominant framework with these two alternatives has been carried, and I believe thoroughly met, by the foregoing discussion in this chapter and, indeed, by the three versions of the whale rescue story recounted in Chapters 3, 5, and 7).

The original expectation had been for Lens I to offer the strongest account of the event, given its apparently ideal embodiment of rationalist tradition/extensionist elements; at best, we hoped that Lenses II and III might prove to have enough relevance to merit their serious consideration as potential supplements. Grounding the critical comparison of the three lenses in this case study, however, revealed results significantly to the contrary: if anything, Story I is the weakest of the three versions of the event, and Story III is surprisingly (given the largely unexplored relevance of Heideggerian phenomenology to environmental philosophy) strong. Ironically, Lens I seems much better suited to accounting for opposition to the event than to justifying it (in part, perhaps, because of the orientation of this tradition toward negative duties rather than positive duties or supererogatory acts, categories into which, from its perspective, the whale rescue probably falls).

Although some of the interviewees were quoted fairly evenly across the three lenses, others gave evidence of holding all three perspectives but were strongly biased toward only one or predominantly used two, usually (but not always) Lenses II and III, largely to the exclusion of Lens I. And although some of the interviewees were quoted in only two stories, none who were quoted more than once failed to appear in Story II. And, interestingly, of the five female interviewees who were quoted at least four times, none accessed Lens I to the exclusion of the others; only one (the least quoted) demonstrated a balanced use of the three perspectives, and all of the remaining four demonstrated a strong preference for Lenses II and III over Lens I.

As additional indicators of the relative strengths of the stories produced by the dominant and alternative frameworks, we might consider the richness, detail, depth, and breadth of either Story II or Story III in comparison with Story I, suggesting that either of the two alternatives seems better able to capture the essence of the participants' experiences than does the dominant model. By and large, participants simply did not adopt the moral point of view and did not cast the issue in the context of extensionism, at least not to a degree equivalent to or exceeding that of the other two approaches. They did, however, experience personal relationships with the whales as a meaningful part of the instigation and implementation of the effort and they did find positive moral value in responsiveness; and they also found the effort a positive confirmation of the best in human nature and the whales a source of wonder.

In summary, Lens I neglects many important elements of the whale rescue as an environmental ethics issue that Lenses II and III, separately and collectively, capture: the alternative lenses are on many levels, as we have seen demonstrated repeatedly, "truer" to the actual experience of human moral agency. And, even more significantly, their fundamental orientation toward the Cartesian worldview, challenging and dismantling it rather than implicitly co-opting it, renders them less susceptible to the detachment from and objectification of the "other," which the dominant tradition reinforces and which are often thought to lie at the heart of anthropocentrism and the practice of environmental exploitation.

It was probably the most incredible part of this whale rescue story: what it generated worldwide. It does say something about us as human beings, doesn't it?

Chapter 9 The Evolution of Moral Philosophy

Let us draw back for a moment and place this study and the development of environmental ethics as we have been exploring it in a larger context. We have examined two general types of transitions here, both occurring within twentieth-century moral philosophy. We have seen interhuman ethical theory critiqued by and supplemented with environmental ethics, and we have seen the dominant tradition of rationalism critiqued by and supplemented with two alternative traditions. And it has been the interlinked social and environmental crises of the past century that in many ways have served as the impetus for both of these transitions.

Environmental ethics is thus distinct from business ethics or medical ethics in at least one significant way: unlike these other newly emerging subdisciplines, environmental ethics is not a field of *applied* ethics but is rather both the product of and a key contributor to the ongoing evolution of moral philosophy. It is not one more arena for the application of given principles but is instead an "offspring" that is fundamentally challenging the very nature of its "parent," regardless of the particular philosophical tradition in question. Thus, extensionism, for example, challenges the deeply ingrained anthropocentric and individualistic assumptions of rationalism; and ecofeminism similarly plays what Karen Warren has called a "transformative" role in feminism, deepening its critique of rationalism, making explicit the interconnections between various forms of oppression, and making "a responsible ecological perspective central to feminist theory and practice."[1] And, although we have not examined them here, there are a host of similarly radical challenges to traditional ways of conceiving the human–environment relationship: deep ecology, social ecology, and various forms of ecotheology to mention only a few. The rethinking of ourselves and of nature and of our place therein that constitutes environmental ethics is a central theme of the twentieth century, one that seems likely to engender significant alterations in the western worldview of the twenty-first century.

Paradigm Shift: Toward an Ecological Worldview

It has become a fairly common device in the history and philosophy of science to categorize the evolution of western thought into three major historical periods, each dominated by a defining worldview: the ancient-medieval period (with its Greco-Roman, Judeo-Christian worldview), the modern period (with its scientific-industrial, Cartesian worldview), and the emerging postmodern period (with its ecological worldview). The Copernican revolution of the sixteenth and seventeenth centuries is generally identified as the transition period between the first and second worldviews, and the Darwinian revolution is often posited as the impetus for the transition from the modern to the postmodern era (with less consensus because in many ways we are still in the process of paradigm shift). In this context, the two transitions within moral philosophy under consideration are clearly components of the broader-scale rethinking currently taking place in the west and are contributing strands in the formation of the emerging western worldview: both are critical of the core assumptions of the modern era (hierarchical dualism, anthropocentrism, limited definitions of rationality and of the self as autonomous, and so on) and both offer alternative conceptualizations that are coming to be seen as characteristic of the postmodern era (a recognition of interdependence and a move toward integration on many levels, systems-thinking, nonanthropocentrism, a validation of experience, and so forth).

Stephen Sterling has arguably written one of the best summaries of the need for and elements of the contemporary transition in the western worldview. His article "Towards an Ecological World View" succinctly captures the essence of the "rising tide of thinking and practice" that is beginning to see in the modern worldview the underlying causes of contemporary environmental and social crises, to challenge these old assumptions, and to replace them with new ways of thinking and new norms for behavior. "Put simply," he writes, the modern worldview bequeathed to the west by the scientific revolution "no longer constitutes an adequate model of reality"; this worldview is secular and mechanistic and materialistic, it hinges on assumptions of separation and dissociation (recall here the hierarchical dualisms of Cartesian rationalism), it defines value in instrumental terms, and it is at best uncertain about the role of ethics in science. "Necessary to the liberation and flowering of scientific inquiry as we have come to know it" and largely successful in rendering us "masters and possessors of Nature," the "schism implied [by this way of thinking] is now at the heart of our contemporary crises."

Although "it is still evolving," Sterling continues, the ecological worldview "contains the seeds of a cultural transformation equal to or surpassing that of the scientific/industrial revolution." Quantum physics and ecology are the defining sciences of this emerging worldview; the former challenges the dissociation of subject and object (the very act of looking at subatomic particles, for example, affects what is observed), the latter insists upon a fundamental and far-reaching awareness of interdependence on all levels, and both are part of "an essential awakening to the interactive character of our relationship with the world." According to the ecological worldview, relationships within systems are central, reality is best thought of as a web of interconnections and interdependencies across both time and space, multiple and often interdisciplinary perspectives are of value, science is understood to have ethical dimensions, and value is conceived of as both instrumental and intrinsic. In short, the new paradigm is one of connection rather than dissociation.[2]

And we increasingly see in the world around us, and in this very study, just such a shift in understanding: interdisciplinary education is on the rise, environmental problems and social problems are seen to be interlinked elements of a larger "problematique," formerly discounted perspectives are sought after and difference is beginning to be valued rather than used to justify exploitation, science and ethics are seen to be inescapably intertwined, qualitative data is legitimated, and humans and nonhumans are understood to be fundamentally interrelated. The western worldview and, within it, moral philosophy moves from anthropocentric to nonanthropocentric theories of value, from a deliberation procedure based on separation of self and other to procedures grounded in interrelatedness, from an understanding of the self as autonomous to a definition of the human as fundamentally embedded and participative, from a dismissal of nonhumans and nature at large as others to be exploited to an appreciation of them as others to be respected and cared for in and of themselves and as interrelated with our own being.

Conceptual Implications of the Ecological Perspective

What, then, are the central implications for moral philosophy of this transition to an ecological worldview? The rise of environmental ethics theorizing itself is perhaps its primary manifestation, and we have examined that development in the evolution of moral philosophy in some depth in Chapters 2, 4, and 6. Let us here take note of three subthemes of environmental ethics that we have yet to make explicit, noting that there are many other examples of the impact of the ecological perspective on moral philosophy.

The Realization of Continuity between Human and Nonhumans

At the heart of the ecological perspective as it traces back to the Darwinian revolution is an understanding of human beings as animals, subject to biological and evolutionary processes, and thus an awareness of the continuity between ourselves and nonhuman animals. Of course, this stands in sharp contrast to both the medieval "Chain of Being" and the modern dualism of human subject ("that which thinks") and nonhuman object ("that which is thought of"). Warren writes that this new understanding of human nature, informed by an awareness of interconnectedness between human and nonhuman, "involve[s] a psychological restructuring of our attitudes and beliefs about ourselves and 'our world' (including the nonhuman world), and a philosophical rethinking of the notion of the self such that we see ourselves as both co-members of an ecological community and yet different from other members of it."[3] Environmental ethics thus involves just such a sorting out of the similarities and differences between ourselves and the other animals and of the implications of this ecological understanding of continuity for moral philosophy.

Animals as subjects
As one example of the consequences of this fundamentally different conceptualization of the human being in relation to the rest of the world, the ecological perspective leads us to consider nonhuman animals as subjects like ourselves. Eliminating this Cartesian dualism by acknowledging the subject-

hood of nonhuman animals, of course, does not necessitate a corollary era-sure of all difference between human as subjects and other animals as sub-jects. Quite the contrary. In fact, acknowledging animals as subjects is the first step toward knowing them as independent individuals, similar in some ways to us and to each other but also ultimately unique. And unlike the Carte-sian worldview, this ecological perspective acknowledges that animals, like ourselves, are in large part defined and shaped by a web of relationships; as Evernden explains, "An animal is not just genes. It is an interaction of genetic potential with environment and with conspecifics. A solitary gorilla in a zoo is not really a gorilla; it is a gorilla-shaped imitation of a social being which can only develop fully in a society of kindred beings."[4]

Anthony Weston proposes several elements on which authentic interac-tion with animals as fellow subjects depends. He writes that "what 'connec-tion' with other animals ultimately requires is a set of practices and com-portments that invites connection, that approaches them as co-inhabitants of a shared world from the start; by taking them seriously as creatures who *have* a point of view, and by . . . paying attention." He suggests that relating to animals as subjects, being open to seeing the world from their perspec-tive, is supported by and in turn engenders refusal to refer to them in objec-tifying "thing" language as "its": " 'he' or 'she' asserts commonality rather than difference, life rather than objecthood, . . . [and] keeps open an aware-ness that we are speaking of genuine creatures in their own rights." The en-vironmental philosophy implications are clear: when we relate to animals as subjects, opening ourselves to the possibility of adopting their perspective and connecting with them as fellow creatures, Cartesian objectification and exploitation are invalidated; "the dismissal, and the death-dealing, are stayed."[5]

And in this light, anthropomorphizing—practiced as the "cop-out" of in-terpretively imposing our own subjective experiences on other beings rather than making the effort to truly experience their own perspective—is thus suggestive of an incomplete, inauthentic attempt to relate to animals as in-dependent subjects. It is important to note, however, that anthropomor-phizing can also be evidence of a sincere attempt to put ourselves in the place of other beings, an attempt that confirms our having overcome the Cartesian view of animals as objects lacking subjective experience but that also attests to our lack of practice in relating to animals as subjects and thus our limited ability to grasp and then express their perspective independently of our own (a shortcoming, it should be noted, that similarly characterizes our efforts to "put ourselves in the shoes of" other people).

Nonhumans as legal entities

And as a second example of the contemporary challenge to the discontinuity model, lawyer and environmental policy theorist Christopher Stone takes the process of extending considerability to nonhumans a step further in his article "Should Trees Have Standing?—Toward Legal Rights for Natural Objects." Noting the parallel between moral standing and legal standing, he calls for the extension of the latter concept to nature as well. "The criteria . . . [for being a] holder of legal rights," he writes, "go towards making a thing count jurally—to have a legally recognized worth and dignity in its own right, and not merely to serve as a means to benefit 'us,'" and he proposes "that we give legal rights to forests, oceans, rivers and other so-called 'natural objects' in the environment—indeed to the natural environment as a whole." As with the extensionist philosophers, Stone sees such a move as the next logical step in the historical process of "making persons of those that were not, in law, always so": children, prisoners, aliens, women, the insane, racial minorities, fetuses, and even corporations, municipalities, nation-states, and ships. Under this proposal, natural objects would have legal standing, meaning that guardians could bring suit on their behalf (not merely on behalf of their users) on the grounds of damages done to them directly (not merely to their users) and that remedies would be addressed directly to their benefit (not merely to the compensation of their users).[6]

An Expansion of Concepts of Value beyond the Model of Rights and Personhood

While critiquing the discontinuity between humans and nonhumans assumed in the modern worldview, the ecological perspective also leads us to question the adequacy and legitimacy of such extension of traditionally human-limited concepts beyond the human realm. As we just saw with Stone's extension of legal standing beyond the human realm, the core assumption of much environmental ethics is that the expansion of the moral community to include at least some nonhumans closely parallels earlier expansions that served to include previously excluded humans; thus not only is moral considerability extended but so are the central concepts of human ethical theory, such as equality and justice and rights. Regan, for example, linking the possession of rights to the possession of inherent value, sees no relevant gap between the human and nonhuman realm that might render the extension of the concept of rights problematic, and thus he claims that "the case for the one [animal rights] is neither any weaker nor any stronger than the care for the other [human rights]."[7]

The ecological perspective, however, questions the extent to which these concepts are translatable from the human to the natural realm. Taylor, for example, does not include the concept of rights in his theorizing, believing that the term adds nothing of significance to our understanding of respect for nature and, in fact, only confuses things by artificially importing a concept that is conceptually limited to the human realm. "Neither plants nor animals can logically be conceived as bearer of moral rights," he states, because they lack the characteristics of personhood on which "the original idea of moral rights . . . in its full, uncompromised meaning as applicable to humans alone rests": they are not (for the most part understood to be) moral agents, they lack the notion of self-respect, they cannot choose whether or not to exercise a given right, and they cannot demand restitution for the infringement of rights.[8]

And, of course, the extension of concepts from human ethical theory becomes even more problematic in an ecocentric approach to environmental ethics, which is even further removed from the original atomistic, human context. Thus, Callicott argues that the notion of the fundamental equality of persons cannot be translated into the natural realm where the value of any given individual (human, animal, plant, or microorganism) can only be defined relative to its contribution to the whole; stating that "the land ethic manifestly does not accord equal moral worth to each and every member of the biotic community," he suggests that it is the distinctness of ecological from social (human) systems that renders the extension of such concepts problematic.[9] Thus, the question to be addressed from the ecological perspective is whether, in Callicott's words, "the elasticity of conventional Western moral theory is limited"[10] by the uniqueness of the political and cultural context in which they first evolved or whether their extension or, interestingly, resistance to their extension or both is merely further confirmation of the pervasiveness within environmental philosophy of the very anthropocentrism it seeks to transcend.

But what form might an alternative conceptualization of value take? Westra suggests that ecologically oriented care signifies "a diffuse respect which requires no specific 'bearers of rights' for its existence . . . [and which] seems to require no specific persons, compacts, or even an understanding of determinant values or interests to justify the obligation we now feel toward . . . the earth."[11] Weston more specifically explores a holistic conception "according to which values are connected in a weblike way" as an alternative to the project of grounding environmental ethics in the intrinsic value of nonhumans. The notion of intrinsic value, he argues, is subtly atomistic in that "value in itself"—deliberately conceived of as a necessary alternative to "value as a means for other ends"—denies interdependence and relatedness (from the ecological perspective, there is no "in itself"), and he posits instead an interdependency model in which multiple values are justified in relation to one another:

> Values may refer beyond themselves without ever necessitating a value which must be self-explanatory. The value of a day's hike in the woods need not be explained either by the intrinsic value of my appreciation of the woods or by the intrinsic value of the woods themselves; instead, both the appreciation and the woods may be valuable for further reasons, the same may be true of *those* reasons, and so forth. Appreciation may be valued . . . partly because it can lead to greater sensitivity to others; but greater sensitivity to others may in turn make us better watchers of animals and storms, and so on. The woods may be valued not only as an expression of freedom and nobility, but also as a refuge for wildlife, and both of these values may in turn be explained by still other, not necessarily human-centered values.[12]

And finally, Holmes Rolston explores the notion of nature itself "as a *source of values*, including our own," positing that "value is not just a human product . . . [in that] we humans . . . [and] much that we value in ourselves, are . . . natural productions of value." Writing that "our place in the natural world necessitates resource relationships, but there comes a point when we want to know how we belong in this world, not how it belongs to us," Rolston suggests a distinction between viewing nature as a *resource* and viewing it as a *source*, exactly the distinction we make in relating to our parents. In this light, "value is a mutifaceted idea with structures that root in natural sources . . . [and] nature is a generative process to which we want to relate ourselves and by this to find relationships to other creatures."[13]

The Attempt to Conceive of Both Individuals *and* Wholes as Loci of Value

Aldo Leopold, arguably the most ecologically oriented of all the authors we have examined here, is generally credited with the impetus toward ecocentrism or, in other words, with the initial challenge to individualistic conceptions of value. At least in his early writings, however, Leopold also posited the revolutionary notion of the considerability of both individuals *and* larger systems: redefining the role of the human being from "conqueror of the land community" to "plain member and citizen," "the land ethic," he wrote, "implies respect for [our] fellow members, and also respect for the community as such."[14] We noted in Chapter 2 that the ecocentric position in extensionism developed the latter emphasis on the considerability of larger wholes, but here we should note also that one consequence of the ecological perspective with its tendency toward integration and its refusal of either-or thinking is an effort to attend to both elements of Leopold's injunction, to understand value as residing in *both* individuals and the larger systems they constitute.

Several ecofeminist authors have hinted at the need to transcend the Cartesian assumption that pits individuals and systems against one another in the "competition" for considerability and for priority of claims. Both individuals and the larger systems they constitute are ultimately defined by relationships and interdependencies, and this weblike view refuses to legitimize hierarchical characterizations of value. Kheel critiques the tendency of both individualistic and ecocentric extensionist approaches to assume that acknowledging value at either level precludes acknowledging it at the other and the inability of either camp to "see that moral worth can exist *both* in the individual parts of nature *and* in the whole of nature of which they are a part." She proposes a reclaiming of the term "holism" from these hierarchical, dualistic associations and a rethinking of the corollary definition of "the 'whole' as composed of discrete individual beings connected by static relationships"; and she advocates a new conception of holism "that perceives nature . . . as comprising individual beings that are part of a *dynamic* web of interconnections." This nondualistic conceptualization "invites us to see value not as a commodity to be assigned by isolated rational analysis, but rather as a living dynamic that is constantly in flux"; in this light, "we cannot postulate that one species or one individual is of greater of lesser value than another."[15]

Similarly, Eric Katz posits a modified holism that takes into account both "the good for the community as a whole and the good for each and every member of the community as an individual." Katz offers the distinction between a "member" of a community, "which exists in its own right *and* as a functioning unit of a community," and a "part" of a larger whole, which has no independent being; thus, in addition to the role animals, plants, and natural objects play in the larger ecosystem, "they also have independent lives and functions." "Natural individuals live and act for and in themselves and as members of a communal system . . . [and thus have] intrinsic value in [themselves] and instrumental value as . . . functioning part[s] of a system," he writes. Katz's ecologically oriented community model of holism thus redefines individuals as individuals-within-systems, acknowledging and balancing both their inherent value as semi-independent beings and their instrumental value as contributors to the larger system, and refuses the exclusive attribution of value to either the individual or the system: "An environmental ethic," he concludes, "must take into account the good for the community as a whole, and the good for each and every member of the community as an individual."[16]

From Moral Monism to Moral Pluralism

And, finally, this exploration of value at the level of both individual and system, in light of the validity granted to multiple perspectives in the ecological worldview, attests to the emergence of a pluralistic approach within moral philosophy (as does, we should note, the attention to context in feminist theory and the understanding of reality as "constructed" in phenomenology). Christopher Stone applies the term "moral monism" to the pervasive assumption within moral philosophy that "the ethicist's task is to put forward and defend a single overarching principle (or coherent body of principles), such as utilitarianism's 'greatest good for the greatest number' or Kant's categorical imperative, and to demonstrate how it (the one correct viewpoint) guides us through all moral dilemmas to the one right solution." The current paradigm shift, however, involves new types of ethical questions and "when one starts to wonder about exotic clients, such as future generations, . . . embryos, animals, . . . trees, robots, mountains, and art works," the adequacy of moral monism becomes problematic and the question arises of whether the same rules and assumptions must apply to all these disparate cases of ethical concern. The alternative, "moral pluralism," "refuses to presume that all ethical activities (evaluating acts, actors, . . . rules, . . . etc.) are in all contexts (in normal interpersonal relationships, across . . . generations, between species) determined by the same features (intelligence, sentience, . . . life) or even that they are subject, in each case, to the same overarching principles (utilitarianism, Kantianism, . . . etc.)." And since each of the varied "conceptual planes" of consideration "brings along its allied constellation of concepts," a pluralistic approach that considers a given dilemma from multiple perspectives avoids the tendency of monism to narrowly define and describe the issue in question and thus produces richer (albeit more complex) analyses.

From this perspective, it is a central issue for the development of environmental ethics whether such theorizing is to subsume existing theory, "mediating all moral questions touching man, beast, and mountain . . . by reference to [an] . . . all encompassing set of principles," or whether it is to "govern man's relations with [nature] alone, leaving intact other principles to govern actions touching humankind."[17] Similarly, with respect to the second of the two transitions in moral philosophy under consideration here, how are we to understand the relationship between the various traditions within western philosophy? This study has hinted at two different general forms such pluralism can take, both between interhuman and environmental ethics and among the three traditions:

Compartmentalization, or Allocating Our Concerns to Different Realms of Moral Theory

The tension over the potential limits of care as discussed in Chapter 4 leads directly to consideration of just this issue in moral philosophy: the failure of both rationalist and care traditions to successfully encompass moral obligation or response to both particular and general others, to both related others and strangers, points (albeit controversially) to the potential for combining the two perspectives in order to "cover all the bases" (as it were). The most obvious and simplistic approach involves dividing the realm of morality into two mutually exclusive sets, with the tradition of rationalism reigning over impersonal interactions and the care tradition over personal ones. This approach has little appeal to most feminist ethicists, however, because it perpetuates the Cartesian dualism of public and private. Meyers probably speaks for most feminist moral philosophers in stating that although "it is a mistake for an ethic of care to jettison the value of justice, and . . . for an ethic of justice to jettison the value of care, . . . none of this entails that . . . these two ethics can be smoothly blended into a single unified approach to moral reflection."[18]

This compartmentalization approach is also being explored in environmental ethics, though not without similar controversy; Kheel, for example, writes that "To dichotomize reality in such a manner . . . [by applying] one set of moral rules . . . to nature, while another applies to society . . . is simply to accept the existing divisions of our society which views itself as separate from (and, in fact, opposed to) nature, rather than simply an extension of it."[19] We noted of Taylor's theory of "respect for nature," for example, that he intends this body of theory to parallel and operate alongside a similar ethical system of "respect for persons" and that he distinguishes yet a third realm of concern (interactions between humans and nonhumans in artificial as opposed to wild settings) subject to "the ethics of the bioculture." Thus three distinct, nonoverlapping sets of theory are posited and we are to divide up the world of ethical concerns accordingly, with recourse to an overarching set of priority principles when the duties specified within one realm come into conflict with those of another.

Callicott has proposed a similar but perhaps less dualistically construed compartmentalization in his attempt to create legitimate and noncontradictory space for the individualistically oriented animal liberation/rights theory and the holistically oriented ecocentric theory. Our human communities are actually "mixed communities" of humans and nonhumans (pets, barnyard animals, and so on), and the domestic animals that are members of this community "ought to enjoy, therefore, all the rights and privileges . . . attendant upon that membership." Wild animals, on the other hand, are not members of this same community but rather are members of the "biotic community," and so different standards and rules apply to our interactions with them. The result, for Callicott, is (roughly) the recognition of two largely distinct ethical systems, both of which arise from an understanding of obligation within community: the animal liberation/rights version of extensionism applies within the domestic realm (the "mixed" human–nonhuman communities) and the ecocentric version applies in the wild. It should be noted here, however, that Callicott, like Taylor, strives to retain the coherence of monism by postulating an overarching structure that adjudicates incommensurabilities, conflicts, and trade-offs among the (otherwise distinct) multiplicity of ethical systems; here the unifying theoretical "umbrella" involves an understanding of ourselves as members of multiple communities and an understanding of these communities as nested and thus conferring obligations of varying "pull" loosely corresponding to the proximity of the community in question.[20]

Complementarity, or Bringing Multiple Approaches to Bear on a Single Realm of Concern

Alternatively, moral philosophy might also take the approach of combining multiple bodies of theory and thus coming to understand a given dilemma in light of different perspectives; this, of course, is the approach taken in this study. Feminist theorists are exploring the compatibility of rationalism and care along just these lines, and attempts to integrate the two perspectives generally begin with Gilligan's claim that "all human relationships . . . can be characterized both in [rationalist tradition] terms of equality and in [care tradition] terms of attachment, and that both inequality and detachment constitute grounds for moral concern."[21] Blum summarizes and expands Gilligan's characterization of the two perspectives as complementary rather than opposed:

> Gilligan does not suggest that care and responsibility are to be seen either as *replacing* impartiality as a basis of morality or as encompassing *all* of morality, as if all moral concerns could be translated into ones of care and responsibility. Rather, Gilligan holds that there is an appropriate place for impartiality, universal principles, and the like within morality, and that a final mature morality involves a complex interaction and dialogue between the concerns of impartiality and those of personal relationship and care.[22]

Gilligan thus acknowledges the legitimacy and relevance of both perspectives and notes that this complementarity is of even greater value when both justice and care orientations are accessed in the course of ethical deliberation, insofar as the two perspectives "do not negate one another but focus attention on different dimensions of the situation . . . [and thus] lead the same situation to be seen in different ways."[23] Meyers writes in this light that "the subject of moral reflection is best construed as a self that draws on empathetic . . . capacities, as well as on impartial rational ones. Equipped in this way, the moral subject is capable of bringing disparate moral themes to bear on a single issue and is capable of profiting from tensions between the moral conclusions these themes suggest."[24]

Manning has proposed an intriguing model of the relationship between justice and care; in her article "Just Caring" she suggests that an ethic of care can, without inconsistency, acknowledge at least three roles for such rationalist concepts as rights and rules:

1. These concepts can be legitimately used as "tools of persuasion to protect the helpless"; thus, "we can reason in the language of rules [and rights] with those . . . [whose] natural sympathies are not engaged by the presence of suffering."

2. The universal concepts of the tradition of rationalism can be used by an ethic of care to ensure "a minimum below which no one should fall and beyond which behavior is morally condemned"; although an ethic of care will not judge "staying above this minimum [as] a sufficient condition for being a morally decent person"—since responding sensitively to others' needs often involves quite a bit more attention, effort, and commitment— appeals to such universal standards can be used effectively to mobilize collective caring actions in defense of "moral minimums."

3. "Rules and rights can also be used to deliberate under some conditions";

for example, "when I am not in direct contact with the objects of care, my actions cannot be guided by the expressed and observed desires of those cared for, and hence I might want to appeal to rules . . . [that] speak to us of what most of us would want as a caring response in a similar situation."

In no case, however, can we allow the rights or rules to become absolutes or ends in themselves or "to distance us from the objects of care" because this would betray the spirit of care as an ethic.[25] Cheney summarizes of this approach to integrating the two perspectives that it, in essence, relies on justice to "approximate the effect of [care's] responsiveness" in certain "special [and] one hopes infrequent" cases in which care alone cannot succeed; thus, "justice is an expedient device for use in situations where, for some reason, we cannot determine what appropriate care consists in, . . . [where] the web-like defining relations of care and responsibility are not in place, . . . or where the situation overwhelms any attempt [at resolution] by means of the established relations."[26]

In the spirit of moral pluralism's complementarity approach, we conclude this study by considering the potential integration of our three lenses. Like it or not, we are products of the tradition of rationalism, and as such we tend to prefer tidy approaches that offer closure; we are more comfortable with the singular answers of monism than with the uncertainty of pluralism, and so in this case we might feel the need at this point to either choose one of our three alternative frameworks as "the best" or develop some sort of "grand unification theory" or overarching framework that unites all three. Again, though, any such effort likely reinforces the very Cartesian assumptions that we have taken such pains to expose and transcend. In this light, then, let us make some tentative suggestions regarding potential integration among our three lenses, especially in terms of their value as complements to one another.

Both in practice and theory, the three lenses are neither mutually exclusive nor necessarily contradictory. In general, study participants spoke from more than one of our three perspectives. Thus, for example, one observer discussed the symbolic value of the whales (Lens I) while at the same time acknowledging that they remain a part of him (Lens II); another related her emotional reaction to and empathy with the whales (Lens II) and then explained her refusal to touch the whales in terms of respect for their Being (Lens III); and still another expressed an obligation to assist the effort grounded in the whales' inherent value (Lens I) and noted the influence of both seeing them with her own eyes (Lens II) and seeing their diminished Being gradually return over the course of the effort (Lens III).

Similarly, we can imagine "real-world" situations in which these different approaches to conceptualizing moral dilemmas and making ethical decisions might well reinforce one another when appealed to or applied in combination. The notion of rationalist concepts establishing a "moral minimum" (protecting those who might otherwise "fall between the cracks" of personal relationships and serving as a banner of sorts to mobilize collective caring) would seem to be equally attractive from the perspective of Lens III, insofar as the largely anthropocentric, "technological" public interpretation might otherwise encounter no barrier to the exploitation of other beings, especially given the gradual pace at which societal norms might be expected to become more "caring." And there is something to be said for speaking a language people understand in one's efforts to educate or change behavior. Given the familiarity of "rights talk," then, it may be that such Lens I concepts are more accessible and meaningful than concepts from the other two lenses; in other words, most people can make sense of "animal rights" by analogy to "human rights," whereas "field of care" and "de-severance" may initially serve as obstacles to communication. On the other hand, however, since Lens II and Lens III concepts derive from actual experience more than from abstract theorizing, it may be that they, rather than Lens I, have a more immediate, intuitive resonance and are thus the better choice for this purpose of effective communication; certainly the claim that a whale is a being "in and of and for himself" and therefore is to be respected as the being he is comes a bit more naturally to the layperson than Kant's "ends in themselves."

Complementary as theory, as well, the traditions-as-lenses focus attention on different aspects of the same issue, each highlighting what others neglect and thus cumulatively adding to our understanding of events like the whale rescue. For example, while Lens I accounted for the initial decision to attempt the rescue (and opposition to that decision), Lenses II and III carried our attention beyond the instigation of the effort to the detailed unfolding of the event and thus offered additional insight and explanation. Similarly, several elements that Lens I was able to account for only as failures of deliberation were treated less dismissively and thus with greater attention from the other perspectives; thus, for example, Lens II took into consideration and meaningfully explained the variable significance of words and pictures as sources of information and motivation, whereas Lens I only noted the media in terms of its role as a conduit for the information on which deliberation depends. Or, again, where Lens I casts understanding or knowledge in terms of reason alone, Lens II offers additional insight by considering emotion in combination with reason as an additional avenue and Lens III adds the perspective of wondering appreciation as yet another route to understanding.

Although the frameworks themselves may not acknowledge the incompleteness inherent in their selecting certain variables as significant to the exclusion of others (Lens I, for example, does not consider emotion to be relevant and therefore sees no incompleteness in omitting it), from the perspective of our critical comparison we are able to recognize the self-limiting nature of the lenses (the fact that each necessarily, and selectively, leaves out pieces of the world in constructing its account of reality) and thus to see them as complementary building blocks that, collectively, help to construct a more complete picture (picking up more, if not most, of the pieces). In simpler terms, whereas Lens I acknowledges that X, Y, and Z did, as a matter of fact, occur, it dismisses Y and Z as irrelevant and focuses attention instead on X; Lens II, on the other hand, dismisses X and Z as irrelevant and explores only Y, whereas Lens III finds X and Y meaningless and directs our attention to Z. Only by combining all three lenses do we come to understand all three variables; and, of course, it very well could be the case that Lens III, for example, finds variable W relevant where Lenses I and II had not even noticed its occurrence.

Addressing this point in the study on which this current work is explicitly modeled, *Essence of Decision: Explaining the Cuban Missile Crisis,* Graham Allison suggests that "the best analysts . . . manage to weave strands of each of the three conceptual models into their explanations," an accomplishment which requires "considerable intuitive powers" and results in "significantly strengthened" analysis.[27] Aldo Leopold may be one of the few in the field of environmental ethics to have blended (albeit without benefit of recognized models) these perspectives in his own thought. In this light, the title "father of environmental ethics" may very well take on a much deeper (and more intriguing) meaning than has been thought. We saw in Chapter 2 how his work is often taken to be representative of extensionism, so here we will merely take note of the presence of both "care" alternatives as well (readers are encouraged to study Leopold in this light in further depth on their own).

It is difficult not to see Leopold's recollection of seeing "a fierce green fire dying in [the] eyes" of a wolf he had just shot and then questioning his long-held predator eradication policy, his characterization of wilderness as "something to be loved and cherished, because it gives definition and meaning to his life," or his justification of selectively cutting birch over pine in terms of an admitted bias for the latter ("I love all trees, but I am in love with pines") as precursors to the care perspective of ecofeminism. Along the same lines, Leopold seems to value the integration of reason and emotion in our ethical thinking, writing that "the evolution of a land ethic is an intellectual as well as emotional process," and he recognizes the significance of particular images in ethical responsiveness, claiming that "we can only be ethical in relation to something that we can see, feel, understand, love, or otherwise have faith in." He repeatedly decries the loss of a felt connection to the land (considering the "spiritual dangers in not owning a farm," for example) and the pervasiveness of an adversarial relationship with nature, and he makes numerous pleas for an awakened consciousness of physical and spiritual interdependence, often grounded in accounts of his own experiences and in appeals to his readers to seek out and learn from such experiences ourselves (to plant a garden and to split wood for a fire, for example).

And turning to the phenomenological perspective, Evernden recounts Leopold's discussion of Wisconsin's monument to the extinct passenger pigeon in light of our unique ability to mourn the loss of a species in his explanation of a Heideggerian understanding of our uniqueness as humans and our consequent responsibility as "shepherds." Leopold models the approach of viewing the world from the perspective of other beings; "curious to deduce [their] state[s] of mind and appetite," he tracks a skunk, a mouse, and a hawk and ponders the significance to each of a January thaw, he advocates "thinking like a mountain" when making wildlife management decisions because "only the mountain has lived long enough . . . [to grasp the] deeper meaning[s]" at stake, and he buries himself in a marsh to witness and share the education of young grebes. Leopold's definition of a true conservationist is strikingly Heideggerian in its emphasis on attitude rather than merely action: "It is a matter of what a man thinks about while chopping, or while deciding what to chop," he writes, and clarifies that "a conservationist is one who is humbly aware that with each stroke he is writing his signature on the face of this land." And the same parallel can be drawn between Heidegger's concern over the consequences of the predominance of revealing as technology and Leopold's concern that "civilization has so cluttered [the] elemental man-earth relation with gadgets and middlemen that awareness of it is growing dim" and his lament that education is no longer moving us "toward soil, [but] away from it." In short, *A Sand County Almanac* is thoroughly pluralistic in its approach, speaking of rights (Lens I), of empathy and affection (Lens II), and of respect and wonder (Lens III); the entire work is a richly diverse collection of Leopold's own stories of living with and learning from the land, a testament to the significance of a pluralistic approach to environmental philosophy and to storytelling.[28]

Particularly in its tendency toward pluralism, then, the emerging ecological worldview refutes the privileging of any one perspective or theoretical framework to the exclusion of others. It critiques the quest for the "one true story," which has long preoccupied the western mind as a misguided corollary to modern dualism and exclusivity, and it nurtures instead a receptiveness to a wide variety of perspectives and stories. And given the environmentally conscious sensitivities of the emerging paradigm, such stories increasingly attend to the defining relationships with nonhumans that constitute so much of our being. It is through the sharing and analysis of these multiple and mutually enriching stories—as we have attempted in this study—that we come to know and appreciate the diverse meanings and possibilities inherent in our fundamental connectedness, whether with past or future, with humans or nonhumans, with friends or strangers, or with whales on the ice.

Appendix: Study Participants and Resources

Interviewees (with 1988 Positions and Interview Dates)

A. S. Micky Becker, executive assistant, Arco Alaska (June 23, 1994)

Charles F. Becker, marketing consultant, Veco Inc. (June 22, 1994)

Arnold Brower Jr., advisor to the Alaska Eskimo Whaling Commission, North Slope Borough administrator (July 11, 1994)

Price Brower, North Slope Borough Search and Rescue (July 8, 1994)

Bonnie Carroll (then Mersinger), White House Aide (June 25, 1994)

Geoff Carroll, biologist, North Slope Borough Department of Wildlife Management (July 11, 1994)

Dennis Covel, Alaska National Guard, Skycrane helicopter crew (June 29, 1994)

Randy Crosby, director, North Slope Borough Search and Rescue (July 8, 1994)

Virginia Crosby, Barrow resident (July 8, 1994)

Mike Doogan, city editor, *Anchorage Daily News* (June 28, 1994)

Dan Fauske, administrative officer, North Slope Borough (July 18, 1994)

Denise Findling, senior photographer, Arco Alaska (June 23, 1994)

Craig George, biologist, North Slope Borough Department of Wildlife Management (July 13, 1994)

Jane Gray, Alaska National Guard (June 29, 1994)

Mike Haller, public affairs officer, Alaska National Guard (July 22, 1994)

Scott Lee, Alaska National Guard, Skycrane helicopter crew (June 29, 1994)

Cindy Lowry, director, Greenpeace Alaska (July 22, 1994)

Charles Mason, Fairbanks photographer (July 6, 1994)

Richard Mauer, reporter, *Anchorage Daily News* (June 30, 1994)

Craig Medred, reporter, *Anchorage Daily News* (June 28, 1994)

Ron Morris, regional director, National Marine Fisheries Service, NOAA (June 27, 1994)

Richard Murphy, photo editor, *Anchorage Daily News* (June 30, 1994)

Bill Roth, photographer, *Anchorage Daily News* (July 19, 1994)

Sherry Simpson, Fairbanks journalist (July 7, 1994)

Ken Waugh, videographer, Arco Alaska (June 23, 1994)

Primary Contributors to Data Collection

Phyllis Clarke, librarian, Tuzzy Public Library, Barrow

Sharon Palmisano, librarian, *Anchorage Daily News*

Mike, Sharon Palmisano's assistant

Librarians, Anchorage Public Library, Alaska Collection

Reference librarians, University of Alaska at Anchorage

Reference and microfilm librarians, University of Alaska at Fairbanks

Microfilm librarians, North Carolina State University

Dr. Tom Albert, senior scientist, North Slope Borough Department of Wildlife Management

Evelyn, Dan Fauske's assistant

Tom Lohman, North Slope Borough Department of Wildlife Management

Leroy Mack, intern, Alaska National Guard

Marc Olson, executive producer, North Slope Borough television studio (and his assistants Natalie Ringland, John, and Robert)

Charles Stalker, pilot, North Slope Borough Search and Rescue

Fran Tate, owner, Pepe's Mexican Restaurant, Barrow

Dee Walker, Veco Inc.

Primary Facilitators (Provided Contacts and Information)

Mark Badger, executive producer, KUAC-TV, University of Alaska at Fairbanks

Colonel Glenn Cassidy, National Guard Bureau, retired

Deborah Fields, National Guard Bureau, environmental affairs

John Manly, press secretary, Alaska Governor's office

Lael Morgan, professor, University of Alaska at Fairbanks

Elise Sereni Patkotak, North Slope Borough Public Information Office

Kim Pigg, development director, KBRW radio, Barrow

Captain Roger Pike, U.S. Coast Guard, retired
Penny Rennick, editor, *Alaska Geographic*
Bill Tobin, Veco Inc., former editor, *Anchorage Times*
Lynn Wilhoite, staffperson, Congressman David Price's office
Staff, Greenpeace Alaska office

Case Study References

Adler, Jerry, Lynda Wright, and Bill White. "Just One Mammal Helping Another."
    Newsweek, October 31, 1988: 74–77.

Bartley, Bruce. "Whale Tale Is Alaska's Top Story." *Fairbanks Daily News-Miner,* De-
    cember 31, 1988–January 1, 1989: A1+.

Behiels, Carmell. Letter. *Barrow Sun,* November 11, 1988: 3.

Berrigan, Christian. Letter. *Anchorage Daily News,* November 7, 1988.

Braham, Howard. Newscast (network and date unknown).

Brower, Arnold, Jr. Videotaped in-house interview. North Slope Borough. Octo-
    ber–November 1988 (exact date unknown).

Brower, Arnold, Sr. Videotaped in-house interview. North Slope Borough. Octo-
    ber–November 1988 (exact date unknown).

Carroll, Bonnie. Personal files.

Casy. Letter to the Alaska National Guard. Mike Haller, personal files.

Chancellor, John. NBC news commentary with Tom Brokaw. October 1988 (exact
    date unknown).

Christine. Letter to the Alaska National Guard. Mike Haller, personal files.

Coates, James. "Ancient Tradition Moved Eskimos to Save Whales." *Anchorage
    Times,* October 31, 1988: A1+.

Cogdill, Ingrid. "Whale Rescue: Case of Misplaced Priorities?" *Fairbanks Daily News-
    Miner,* October 20, 1988: 8.

Colen, B. D. "Whose Rescue Was It?" *Health,* February 1989: 86.

Craig. Letter to the Alaska National Guard. Mike Haller, personal files.

Crane, Barbara. "Operation Breakout." *DMVA Impact,* January 1989: 6.

Diechilman, Cheryl. Letter. *Los Angeles Times,* October–November 1988 (exact date
    unknown).

Dorfman, Andrea. "Free at Last! Bon Voyage!" *Time,* November 7, 1988: 130.

Dow, David. CBS newscast. October 1988 (exact date unknown).

Dunlap-Skohl. Cartoon. *Anchorage Daily News,* October 20, 1988.

Dye, Lee. "Eager Whales Free, Head for the Open Sea." *Los Angeles Times,* October
    27, 1988: D6+.

———. "Whale Rescue: Helping Hand Grows into Arctic Blitzkrieg." *Los Angeles
    Times,* October 24, 1988: 1+.

———. "The Whales of October: Three Stayed for Dinner." *Los Angeles Times,* Oc-
    tober 22, 1988: F11+.

Epler, Patti. "Beluga Rescue Is 'No Big Deal.'" *Anchorage Daily News*, October 24, 1988: A1+.

Epstein, James M. Letter. *Los Angeles Times*, October–November 1988 (exact date unknown).

Fauske, Dan. Videotaped in-house interview. North Slope Borough. October–November 1988 (exact date unknown).

Foster, Mary Bernardine. Letter. *Los Angeles Times*, October–November 1988 (exact date unknown).

Fuller, Chet. "Whale Rescue: The Drama Involved Overshadowed Other Priorities." *Atlanta Journal and Constitution*, October 31, 1988: A11.

Gordon, David G. "Following the Gray Whales, From Baja to the Bering Sea." *Alaska Airlines Magazine*, January 1992: 64–67.

Gorner, Peter. "Whales." *Chicago Tribune*, October 30, 1988: G1+.

"Group Founder Says Rescue 'Hypocritical.'" *Fairbanks Daily News-Miner*, October 27, 1988.

Gumbal, Bryant. *NBC Today Show* (date unknown).

Hadley, James R. Letter. *Anchorage Daily News*, November 8, 1988.

Haller, Mike. Personal files.

Heath. Letter to the Alaska National Guard. Mike Haller, personal files.

"Heroic Measures for Gentle Giants." *U.S. News & World Report*, October 31, 1988: 12.

Hess, Bill. *Uiniq: The Open Lead* 3.3 (Fall 1988). Gray Whale Rescue, Collector's Edition.

Hulen, David. "Barrow Residents May Kill Trapped Whales." *Anchorage Daily News*, October 15, 1988: A1+.

Hunt, Bill. "Expected Book on Barrow Whale Rescue Surfaces." Review of *Freeing the Whales: How the Media Created the World's Greatest Non-Event*, by Tom Rose. *Anchorage Daily News*, January 7, 1990: F14.

Hunt, Joe. "Mortality Play Entraps World in Its Emotion." *Anchorage Times*, October 19, 1988: A1+.

———. "Whale of a Tale." *Anchorage Times*, November 27, 1988: B11+.

Jacobs, Joanne. "Nature Should Be Allowed to Take Its Course with Alaska's Gray Whales." *Anchorage Times*, October 25, 1988: B7.

John. Letter to the Alaska National Guard. Mike Haller, personal files.

Josie. Letter to the Alaska National Guard. Mike Haller, personal files.

J. S. Letter to the Alaska National Guard. Mike Haller, personal files.

Kelley. Cartoon. *San Diego Union*, October–November 1988 (exact date unknown).

Kilpatrick, James. "The Rescue of Gray Whales and Us." *Anchorage Times*, November 2, 1988.

Kunen, James, and Maria Wilhelm. "To Save the Whales." *People Weekly*, November 7, 1988: 60–63.

Lefevre, Greg. CNN newscast. October 1988 (exact date unknown).

Letter to the Alaska National Guard. Mike Haller, personal files (child's name unknown).

Linden, Eugene. "Helping Out, Putu, Siku and Kanik." *Time,* October 31, 1988: 76–77.

Matumeak, Warren. Videotaped press conference. North Slope Borough. October 1988. Reading a statement by Mayor George Ahmaogak (exact date unknown).

Mauer, Richard. "As Alaska Whale Rescue Is Pressed, Biologists Question Efforts' Wisdom." *New York Times,* October 24, 1988: 15.

———. "Path Cleared in Ice, Whales Swim Free." *New York Times,* October 27, 1988: A16.

———. "Reluctant Whales Coaxed Closer to Sea." *New York Times,* October 28, 1988: A16.

———. "Soviet Ships Drive a Path through Ice." *Anchorage Daily News,* October 26, 1988: A1+.

———. "Unlikely Allies Rush to Free Three Whales." *New York Times,* October 18, 1988: A18.

———. "Whales Free, Swim toward Open Sea." *Anchorage Daily News,* October 27, 1988: A1+.

———. "Two Whales Apparently Escape to Sea; Rescue Effort Called Success." *New York Times,* October 29, 1988: A6.

McDonald, Bill. "Tide Carries Save-the-Whales Group." *Bridgeport Post* [Connecticut], July 25, 1988: G12+.

McGraw, Carol. "Warmth of Celebrity Status Greets Scientist Who Rescued Whales." *Los Angeles Times,* November 20, 1988: B10.

Medred, Craig. "Saving Endangered Species Takes More Than Cutting Ice Holes." *Anchorage Daily News,* October 31, 1988: B1+.

———. "Soviets Heading to Barrow." *Anchorage Daily News,* October 24, 1988: A1+.

———. "Whale Wait Drags On." *Anchorage Daily News,* October 19, 1988: A1+.

———. "Whale Watch: Why?" *Anchorage Daily News,* October 24, 1988: G1+.

———. "When the Pack Fed on Whales." *Anchorage Daily News,* October 1, 1989: K1+.

Miller, K. C. Letter. *Barrow Sun,* November 11, 1988: 2.

Mitchell, Henry. "With Whale Rescue We Show We Are Part of Nature." *Anchorage Daily News,* October 25, 1988: D7.

Monica. Letter to the Alaska National Guard. Mike Haller, personal files.

Morris, Ron. Newscast. October 1988 (network and date unknown).

———. Videotaped press conference. North Slope Borough. October 1988 (exact date unknown).

Murphy, Karen. Letter. *Los Angeles Times,* October–November 1988 (exact date unknown).

Nelson, Lisa. ABC newscast. Interview with Gary Shepard. October 1988 (exact date unknown).

Nightingale, Suzan. "In a Desperate World, We Help Whom We Can—Even if It's a Whale." *Anchorage Daily News,* October 19, 1988.

Odom, Ben. Videotaped in-house interview. Arco Alaska (date unknown).

O'Leary, Jeremiah. "Reagan Bolsters Alaskan Whale-Rescue." *Washington Times,* October 19, 1988: 19.

Oliver, Don. NBC newscast. October 1988 (exact date unknown).

Passage to Freedom. Narr. Doug McDonnel. Writ. and prod. Dan Sexton and Doug McDonnel. Dir. Jim Lutton. Exec. prod. Ron Lorentzen. KPIX-TV, 1988.

Postman, David. "Whales Slowing Down." *Anchorage Daily News,* October 17, 1988: A1+.

Postman, David, and Craig Medred. "Hunter's Odd Discovery Turns World's Eyes toward Barrow." *Anchorage Daily News,* October 23, 1988: A1+.

Raspberry, William. "Lessons of the Whales." *Washington Post,* October 28, 1988: A23.

Rather, Dan. CBS newscast. October 1988 (exact date unknown).

Reagan, Ronald. Letter to the rescue participants. Bonnie Carroll, personal files.

Reeves, Randall R. "Tales of Leviathan." *Natural History,* March 1989: 76–81.

Reid, Sean. "Whale Rescue: The World Watches." *Alascom Spectrum,* November 1988: 2–5.

Rosenblatt, Roger. "Looking at Them Looking at Us." *U.S. News & World Report,* November 7, 1988: 8.

Ross-Flanigan, Nancy. "Human Interference with Nature: Right or Wrong?" *Anchorage Times,* December 11, 1988: C10.

Salminen, David A. Letter. *Anchorage Daily News,* October 29, 1988.

Sargent, Ben. Cartoon. *Austin American Statesman,* October–November 1988 (exact date unknown).

"Saving Whales Simply Madness on the Arctic Ice." Editorial from *Baltimore Sun. Anchorage Daily News,* October 26, 1988.

Scheffer, Victor B. "Introduction: Alaska's Whales." *Alaska Geographic,* 5.4 (1978): 5–14.

Shabecoff, Philip. "What Three Whales Did to the Human Heart." *New York Times,* November 6, 1988: D11.

Shain. Letter to the Alaska National Guard. Mike Haller, personal files.

Shasta. Letter to the Alaska National Guard. Mike Haller, personal files.

Shaw, Bernard. CNN newscast. October 1988 (exact date unknown).

Shepard, Gary. ABC newscast. October 1988 (exact date unknown).

Simpson, Sherry. "Icebreakers Smashing toward Whales." *Washington Post,* October 26, 1988: A3.

———. "Lessons from Whale Rescue." *Fairbanks Daily News-Miner,* November 6, 1988: H7+.

———. "Rescue May Top $1 Million." *Fairbanks Daily News-Miner,* October 29, 1988: 1+.

————. "Whale Drama Taking Toll on Eskimos." *Fairbanks Daily News-Miner,* October 25, 1988: 1+.

————. "Whale of a Story Leads to 'Icepack Journalism.'" *Fairbanks Daily News-Miner,* October 22, 1988: 3.

————. "Three Whales and a Juggernaut." *Washington Post,* October 21, 1988: C12+.

Staine, Michael. Letter. *Anchorage Daily News,* November 7, 1988.

Stevens, Ted. Letter to President Ronald Reagan. November 18, 1988. Bonnie Carroll, personal files.

Stolfus, Kecia. Letter. *Los Angeles Times,* October–November 1988 (exact date unknown).

Synwolt, David. Letter. *Los Angeles Times,* October–November, 1988 (exact date unknown).

Tetpon, John. "It's a Big Story, OK, But Why?" *Anchorage Daily News,* October 20, 1988: A1+.

Underwood, Nora. "The Whales of Alaska." *Macleans,* October 31, 1988: 46–48.

Wasserman. Cartoon. *Boston Globe,* October–November 1988 (exact date unknown).

Weaver, Howard. "Why All the Hullabaloo about These Whales?" *Anchorage Daily News,* October 23, 1988.

Webb, Edwin, Jr. Letter. *Anchorage Daily News,* October 29, 1988.

"Whale of a Tale." *Los Angeles Times,* October 29, 1988.

"Whale Rescue Captured World's Imagination." Editorial. *Fairbanks Daily News-Miner,* October 23, 1988: A4.

"What's in a Name? Cash." *The Economist,* October 29, 1988: 31.

Will. Letter to the Alaska National Guard. Mike Haller, personal files.

Williams, Ted. "Circus Whales." *Audubon,* March 1989: 16–23.

Wohlforth, Charles P. "Whale Rescue in Works." *Anchorage Daily News,* October 16, 1988: A1.

Notes

The quotes and short whale stories that open Parts I through IV and "Whale Stories" I, II, and III in Chapters 3, 5, and 7 are drawn from the comments of interviewees and the various documents related to the rescue as listed in the Appendix.

Introduction

1. Robert Coles, *The Call of Stories: Teaching and the Moral Imagination* (Boston: Houghton Mifflin, 1989), 204–5, 30.

2. Albert Gore Jr., address to the Democratic National Convention, July 16, 1992.

3. Graham T. Allison, *Essence of Decision: Explaining the Cuban Missile Crisis* (New York: HarperCollins Publishers, 1971), v.

4. Peter Railton, "Alienation, Consequentialism, and the Demands of Morality," *Philosophy and Public Affairs* 13.3 (1984): 134, 135.

5. Kenneth Maly, "Earth-Thinking and Transformation," in *Heidegger and the Earth,* ed. Ladelle McWhorter (Kirksville, MO: Thomas Jefferson University Press, 1992), 11.

6. Annette Baier, *Postures of the Mind* (Minneapolis: University of Minnesota Press, 1985), 235, 236.

7. It is important, therefore, that the researcher acknowledge her own biases and unique perspectives and that she take explicit note of how they are affecting the course of the study, without, of course, allowing them the free rein that would interfere with the process of "getting inside" another's perspective. In this study, I admitted up front my qualms with the theoretical adequacy of the dominant tradition and my expectations with regard to the relevance of the two alternatives; that these alternatives "ring true" in my experience is one of several presuppositions undoubtedly carried into this study. And, second, I basically supported the rescue effort as the appropriate course of action, and I am not under the illusion that readers are unaware of that fact. I did not, however, go out of my way to announce my personal opinion during the course of the interviews because I did not want to set in motion the tendency for interviewees to "tell you what they think you want to hear"; when asked, I generally acknowledged the legitimacy of both supporting and opposing positions and explained my own leanings in a way that, I am quite confident, dissuaded no one from expressing his or her own opinion. (Although this is a major concern among qualitiative researchers, in my experience egos were not so fragile nor opinions so easily swayed that my expression of approval for the effort could have made anyone wary of expressing his or her own disapproval.)

Chapter 1

1. Although environmentalists and the public at large often view Alaska as nature's last stand in this country—as the one place where human influence has yet to alter the face of the land for the worse—and thus feel strongly about environmental preservation when these issues arise, many Alaskans would note that such issues, in fact, loom larger in minds of people living in the Lower 48: among citizens and policymakers who tend to view the state as one big "National Park" rather than as a place where people actually live and work and who thus tend to be overly sensitive and reactionary in the area of environmental regulation.

2. The bowhead hunt of the Inupiat Eskimos is, of course, a highly controversial practice and a ripe candidate for discussion in the realm of environmental ethics; such discussion, however, is beyond the range of this work. The author invites readers to explore in their own minds the ethical considerations on all sides of this characteristically complex environmental ethics issue; as an exercise, consider the issue briefly before reading the following pages and then again at the conclusion of the text. Hopefully the philosophical skills developed and exercised in the process of thoughtful reading will deepen the concluding analysis of this issue and build the reader's confidence in his or her own ethical reasoning abilities. A word of warning, however: the reader, having gone through this process and having become familiar with the ethical perspectives considered herein, should not expect the issue to appear more clear-cut or the answer obvious; a common consequence of such study is that ethical questions come to be seen in light of their true complexity rather than in black-and-white terms.

3. Joseph R. DesJardins, *Environmental Ethics: An Introduction to Environmental Philosophy* (Belmont, CA: Wadsworth Publishing Co., 1993), viii.

Chapter 2

1. Immanuel Kant, *Groundwork of the Metaphysic of Morals,* trans. H. J. Paton (1785; New York: Harper & Row, 1964), 57, 59, 60.

2. Christopher Stone, "Should Trees Have Standing?—Toward Legal Rights for Natural Objects," in *People, Penguins, and Plastic Trees: Basic Issues in Environmental Ethics,* ed. Donald VanDeVeer and Christine Pierce (Belmont, CA: Wadsworth Publishing Co., 1986), 84.

3. Peter Singer, *Practical Ethics* (New York: Cambridge University Press, 1979), 11.

4. Richard Tarnas, *The Passion of the Western Mind: Understanding the Ideas That Have Shaped Our World View* (New York: Ballantine Books, 1991), 4.

5. Ibid., 34.

6. Ibid., 22.

7. William Frankena, *Ethics* (Englewood Cliffs, NJ: Prentice-Hall, 1973), 1–11.

8. Plato, *Crito,* in *Collected Dialogues of Plato,* ed. Edith Hamilton and Huntington Cairns, Bollingen Series LXXI (New York: Bollingen Foundation, 1961), 29–39.

9. Tarnas, *Passion of the Western Mind,* 277–78.

10. Jeremy Bentham, *An Introduction to The Principles of Morals and Legislation* (1789; New York: Hafner Publishing Co., 1948), 1.

11. Ibid., 2.

12. John Stuart Mill, *Utilitarianism,* ed. George Sher (1863; Indianapolis: Hackett Publishing Co., 1979), 7.

13. Bentham, *Principles of Morals,* 29–30.

14. Kant, *Groundwork of the Metaphysic,* 61, 67–68, 70–71, 95–96.

15. John Rawls, *A Theory of Justice* (Cambridge: Harvard University Press, 1971), 136–37.

16. Bentham, *Principles of Morals,* 311 (footnote to paragraph IV).

17. Bryan Norton is the leading voice of opposition to nonanthropocentric readings of Leopold. Offering a more pragmatic interpretation, Norton grounds Leopold's call for a sense of obligation to restrict our exploitation of the environment in human concerns: in anthropocentrism that extends to include future generations and that is informed by awareness of ecological interdependence.

18. Aldo Leopold, *A Sand County Almanac, with Essays on Conservation from Round River* (1949; 1953; New York: Ballantine Books, 1966), 237–40, 244–45, 262.

19. Peter Singer, *Animal Liberation: A New Ethics for Our Treatment of Animals* (New York: Avon Books, 1975), ix.

20. Ibid., 232.

21. Ibid., 11, 176–79.

22. Peter Singer, "Not for Humans Only: The Place of Nonhumans in Environmental Issues," in *Ethics and Problems of the 21st Century,* ed. Kenneth Goodpaster, Kenneth Sayer, and Peter Singer (Notre Dame, IN: University of Notre Dame Press, 1979), 194 (emphasis added).

23. Peter Singer, "Animal Liberation," in *People, Penguins, and Plastic Tress: Basic Issues in Environmental Ethics,* ed. Donald VanDeVeer and Christine Pierce (Belmont, CA: Wadsworth Publishing Co., 1986), 25.

24. Singer, "Not for Humans Only," 196.

25. Singer, *Liberation,* 7. The term "speciesism" is usually associated with Singer but was actually first coined by Richard Ryder. See Ryder, "Experiments on Animals," in *Animals, Men and Morals: An Enquiry into the Maltreatment of Non-Humans,* ed. Stanley Godlovitch, Roslind Godlovitch, and John Harris (New York: Taplinger Publishing Co., 1972).

26. Peter Singer, "All Animals are Equal," in *Animal Rights and Human Obligations,* ed. Tom Regan and Peter Singer (Englewood Cliffs, NJ: Prentice-Hall, 1976), 155.

27. Singer, *Liberation,* 21.

28. Ibid., 255.

29. Tom Regan, "The Case for Animal Rights," in *People, Penguins, and Plastic Tress: Basic Issues in Environmental Ethics,* ed. Donald VanDeVeer and Christine Pierce (Belmont, CA: Wadsworth Publishing Co., 1986), 32 (emphasis added).

30. Ibid., 37.

31. Ibid.

32. Tom Regan, *The Case for Animal Rights* (Berkeley: University of California Press, 1983), 240–41, 245.

33. Ibid., 56–81, 262.

34. Ibid., 248–49, 262.

35. Ibid., 324.

36. Regan, "Case," 38–39.

37. Regan, *Case,* 246 (emphasis added).

38. Ibid., 31.

39. Paul W. Taylor, *Respect for Nature: A Theory of Environmental Ethics* (Princeton, NJ: Princeton University Press, 1986), 3, 13–19, 53, 117, 172–79, 184–97, 259, 264–81, 292–93, 306–8, 312–13.

40. Kenneth Goodpaster, "On Being Morally Considerable," *Journal of Philosophy* 75.6 (1978): 309–11 316, 319–20, 323–24.

41. Singer, "Not for Humans Only," 203, 205.

42. Regan, *Case,* 362, 395.

43. Taylor, *Respect for Nature,* 69, 70.

44. J. Baird Callicott, *In Defense of the Land Ethic: Essays in Environmental Philosophy* (Albany: State University of New York Press, 1989), 3–4. Callicott does not consider himself an extensionist and, in fact, goes to great lengths to distinguish his de-

velopment of the land ethic from the approach of animal liberation/rights. He deliberately disavows the approach of modifying traditional utilitarian or deontological
theory as insufficiently critical of "cherished first principles" (20); and where he
draws on the philosophical canon, his allegiance is to the affective orientation of
Hume, suggesting, as does his emphasis on ecological context and biotic relationships, an affinity for our second tradition. Nevertheless, he retains (with Leopold)
the assumption that the "next step" in the evolution of ethics involves "the extension
of direct ethical considerability from people to nonhuman natural entities," (15) and
his work embodies many of the underlying assumptions of the tradition of rationalism in western moral philosophy. Thus, although the larger project remains, the label "extensionism" is admittedly something of an uneasy fit here; given the scale of
our analysis here and a broader definition of the term than Callicott generally uses
(referencing considerability per se rather than particular bodies of theory), his positioning here is perhaps not inappropriate.

45. Leopold, *Sand County Almanac*, xviii–xix, 117, 210, 212, 239.

46. Ibid., 240.

47. Ibid., 262.

48. Callicott, *In Defense of the Land*, 21, 28.

49. Leopold, *Sand County Almanac*, 240.

50. Callicott, *In Defense of the Land*, 21.

51. Ibid., 21, 23–25, 37.

52. Ibid., 23.

53. Regan, *Case*, xii.

54. Taylor, *Respect for Nature*, 9, 24.

55. Singer, *Liberation*, x–xi, 255.

56. Regan, *Case*, 198. Ironically, one of the best examples of the devaluation of
emotion in this literature is a passionate and moving paragraph at the conclusion of
one of Regan's later articles; after devoting several articles and *The Case for Animal
Rights'* four hundred pages to a rigorous intellectual defense of his ethical theory, Regan finally gives voice to his own deeply emotional and empathetic perspective on
animal suffering in our society, but he allows less than two vertical inches of text for
these—by implication—less important albeit resonant and powerful, words: "As for
the passion: There are times, and these are not infrequent, when tears come to my
eyes when I see, or read, or hear of the wretched plight of animals in the hands of humans. Their pain, their suffering, their loneliness, their innocence, their death.
Anger. Rage. Pity. Sorrow. Disgust. The whole creation groans under the weight of
the evil we humans visit upon these mute, powerless creatures. It *is* our heart, not
just our head, that calls for an end" ("Case," 39).

57. Taylor, *Respect for Nature*, 27, 91, 169–70.

58. Ibid., 178, 179.

59. Regan, *Case*, 399.

Chapter 3

1. Extensionism does not posit that we are obliged to intervene in nature and end all suffering, although Singer and Regan have been (mis)interpreted along these lines. The point here is that the suffering of animals is deemed morally relevant, not that it imposes a duty on us to eliminate or reduce the experience of pain; the tradition would probably consider addressing such suffering to be a supererogatory act, one that is "above and beyond the call of duty." And, of course, many authors would view this rationale as illegitimate grounds for violating our duties of noninterference.

Chapter 4

1. Marti Kheel, "The Liberation of Nature: A Circular Affair," *Environmental Ethics* 7 (Summer 1985): 144.

2. Eve Browning Cole, *Philosophy and Feminist Criticism: An Introduction* (New York: Paragon House, 1993), 44.

3. Diana Meyers, "Moral Reflection: Beyond Impartial Reason," *Hypatia* 8.3 (Summer 1993): 41, 42.

4. Several philosophers from the ancient Greeks to the present, specifically Aristotle and David Hume, have engaged in theorizing that bears explicit resemblance to the care tradition, and their work can thus be examined as its mainstream intellectual seed-bed. Unlike their dominant tradition colleagues, Aristotle and Hume saw actual human experience as the key to understanding morality and as the most secure foundation for ethical theory, and they denied the usefulness of abstract theorizing; it is these crucial assumptions that ally their work, in many ways, with the feminist perspective on moral philosophy (although it must also be noted that, in many other ways, these two men shared their colleagues' disdain for women as lesser human beings and thus cannot actually be considered as forerunners of feminism). In their Introduction to *Women and Moral Theory* (Totowa, NJ: Rowman & Littlefield, 1987, 8), in fact, Diana T. Meyers and Eva Feder Kittay acknowledge the intellectual debt feminist philosophy owes these earlier, and in this respect revolutionary, thinkers by summarizing the key elements of Aristotle and Hume's thought that laid the groundwork for later development by feminist theorists:

> For Aristotle, moral deliberation must determine the right thing to do, at the right time, in the right place, to the right person, in the right way . . . , moral judgment springs from a moral character attuned to circumstantial and contextual features [and] is not the product of an abstract concept of the Good . . . , [and] the social embeddedness of the individual [is important]. . . . In a related vein, Hume's ethics are grounded in emotion and personal concern. Hume argued that reason itself could not move us to act morally, but that our ethical life is guided by moral sentiments. [And], again, attention to relationships is prominent. . . . [Thus], the interest in alternatives to a deductive, calculative approach

to moral decision-making, with its strong emphasis on individual autonomy, may be traced back to Aristotle and Hume.

5. Jim Cheney, "Eco-Feminism and Deep Ecology," *Environmental Ethics* 9 (Summer 1987): 116, quoting Ynestra King, "The Ecology of Feminism and the Feminism of Ecology," *Harbinger: The Journal of Social Ecology* 1 (1983): 16.

6. Karen Warren, "Feminism and Ecology: Making Connections," *Environmental Ethics* 9 (Spring 1987): 4–5.

7. There are at least four primary versions of feminist theory: liberal, Marxist, radical, and socialist. And there is a similar diversity among ecofeminism, ranging from Wiccan/neopagan ecofeminists to feminists for animal rights. "Care" is one strand of feminist theory and, correspondingly, one strand of ecofeminist theory; although there are similarities among all versions of feminism and all versions of ecofeminism, the discussion here is not intended to encompass any feminist thought beyond that embodied in the care tradition.

8. Karen Warren, "The Power and the Promise of Ecological Feminism," in *The Environmental Ethics and Policy Book: Philosophy, Ecology, Economics,* ed. Donald VanDeVeer and Christine Pierce (Belmont, CA: Wadsworth Publishing Co., 1994), 275.

9. Cole, *Philosophy and Feminist Criticism,* 107.

10. Warren, "Power," 274.

11. Warren, "Feminism and Ecology," 8.

12. Cole, *Philosophy and Feminist Criticism,* 2.

13. Val Plumwood, "Nature, Self, and Gender: Feminism, Environmental Philosophy, and the Critique of Rationalism," *Hypatia* 6.1 (Spring 1991): 3.

14. Cole, *Philosophy and Feminist Criticism,* 44.

15. Plumwood, "Nature, Self, and Gender," 11, 17.

16. Warren, "Power," 267–69.

17. Ibid., 267.

18. Plumwood, "Nature, Self, and Gender," 7.

19. Ibid., 3.

20. Kheel, "Liberation of Nature," 137.

21. Plumwood, "Nature, Self, and Gender," 6.

22. Kheel, "Liberation of Nature," 139, 140.

23. Caroline Whitbeck, "A Different Reality: Feminist Ontology," in *Beyond Domination: New Perspectives on Women and Philosophy,* ed. Carol C. Gould (Totowa, NJ: Rowman & Allanheld, 1983), 66, 68–69, 72.

24. Deborah Slicer, "Your Daughter or Your Dog? A Feminist Assessment of the Animal Research Issue," *Hypatia* 6.1 (Spring 1991): 112.

25. Warren, "Power," 275–76.

26. Plumwood, "Nature, Self, and Gender," 17.

27. Mary Midgley, *Animals and Why They Matter* (Athens: University of Georgia Press, 1983), 12.

28. Warren, "Feminism and Ecology," 14.

29. Midgley, *Animals,* 42.

30. Robin Morgan, "Metaphysical Feminism," in *The Politics of Women's Spirituality,* ed. Charlene Spretnak (Garden City, NY: Anchor Press, 1982), 387.

31. Kheel, "Liberation of Nature," 141–44.

32. Stephanie Lahar, "Ecofeminist Theory and Grassroots Politics," *Hypatia* 6.1 (Spring 1991): 36.

33. Roger King, "Caring about Nature: Feminist Ethics and the Environment," *Hypatia* 6.1 (Spring 1991): 86.

34. Cole, *Philosophy and Feminist Criticism,* 61, 62.

35. Plumwood, "Nature, Self, and Gender," 19.

36. Warren, "Power," 276.

37. Ibid., 275.

38. Midgley, *Animals,* 22 (emphasis added).

39. Warren, "Feminism and Ecology," 12.

40. Warren, "Power," 276.

41. Ibid., 275, quoting Cheney, "Eco-Feminism and Deep Ecology," *Environmental Ethics* 9 (Summer 1987): 122.

42. Warren, "Power," 276.

43. King, "Caring about Nature," 79.

44. Seyla Benhabib, "The Generalized and the Concrete Other," in *Ethics: A Feminist Reader,* ed. Elizabeth Frazer, Jennifer Hornsby, and Sabina Lovibond (Oxford: Blackwell Publishing, 1992), 280.

45. Margaret Urban Walker, "Moral Understandings: Alternative 'Epistemology' for a Feminist Ethics," in *Explorations in Feminist Ethics,* ed. Eve Browning Cole and Susan Coultrap-McQuin (Bloomington: Indiana University Press, 1992), 166, 167.

46. Warren, "Power," 276.

47. Ibid.

48. Kheel, "Liberation of Nature," 145.

49. Warren, "Feminism and Ecology," 19.

50. Walker, "Moral Understandings," 169.

51. Cole, *Philosophy and Feminist Criticism,* 48.

52. Carol Gilligan, "In a Different Voice: Women's Conceptions of Self and of Morality," *Harvard Educational Review* 47.4 (November 1977): 511.

53. Carol Gilligan, *In a Different Voice: Psychological Theory and Women's Development* (Cambridge: Harvard University Press, 1982), 74.

54. Ibid., 160.

55. Ibid., 33.

56. Cheney, "Eco-Feminism," 120, quoting Carol Gilligan, "The Conquistador and the Dark Continent: Reflection on the Psychology of Love," *Daedalus* 113 (1984): 93.

57. Meyers and Kittay, Introduction to *Women and Moral Theory,* 10.

58. Gilligan, *Different Voice*, 28, 29, 80.

59. Carol Gilligan, "Remapping the Moral Domain: New Images of Self In Relationship," in *Mapping the Moral Domain: A Contribution of Women's Thinking to Psychological Theory and Education*, ed. Carol Gilligan, Janie V. Ward, Jill McLean Taylor, and Betty Bardige (Cambridge: Harvard University Press, 1988), 7.

60. Carol Gilligan, "Moral Orientation and Moral Development," in *Women and Moral Theory*, ed. Eva Feder Kittay and Diana T. Meyers (Totowa, NJ: Rowman & Littlefield, 1987), 23–24.

61. Gilligan, "Different Voice," 509.

62. Cole, *Philosophy and Feminist Criticism*, 107.

63. Whitbeck, "A Different Reality," 79, 80.

64. Rita Manning, "Just Caring," in *Explorations in Feminist Ethics*, ed. Eve Browning Cole and Susan Coultrap-McQuin (Bloomington: Indiana University Press, 1992), 45, 47.

65. Meyers, "Moral Reflection," 24–26, 30–31.

66. Warren, "Power," 271–73.

67. Carol J. Adams, "Ecofeminism and the Eating of Animals," *Hypatia* 6.1 (Spring 1991): 128–29, 134.

68. Deane Curtin, "Toward an Ecological Ethic of Care," *Hypatia* 6.1 (Spring 1991): 69–70.

69. Slicer, "Your Daughter," 117–19, 121.

70. Ibid., 121.

71. Ibid.

72. Ibid.

73. Plumwood, "Nature, Self, and Gender," 12.

74. Ibid., quoting John Seed, Joanna Macy, Pat Fleming, and Arne Naess, *Thinking Like a Mountain: Towards a Council of All Beings* (Philadelphia: New Society Publishers, 1988), 36.

75. Plumwood, "Nature, Self, and Gender," 15, quoting Warwick Fox, "Deep Ecology: A New Philosophy for Our Time?" *The Ecologist* 14 (1984): 60.

76. Cheney, "Eco-Feminism," 126–27.

77. Plumwood, "Nature, Self, and Gender," 13, 14.

78. Ibid., 14, quoting Jean Grimshaw, *Philosophy and Feminist Thinking* (Minneapolis: University of Minnesota Press, 1986), 182–83.

79. Plumwood, "Nature, Self, and Gender," 20.

80. Joan C. Tronto, "Beyond Gender Difference to a Theory of Care," in *An Ethic of Care: Feminist and Interdisciplinary Perspectives*, ed. Mary Jeanne Larrabee (New York: Routledge, 1993), 249–50.

81. Lawrence A. Blum, "Gilligan and Kohlberg: Implications for Moral Theory," in *An Ethic of Care: Feminist and Interdisciplinary Perspectives*, ed. Mary Jeanne Larrabee (New York: Routledge, 1993), 50.

82. Gilligan, "Moral Orientation," 24, 32.

83. Walker, "Moral Understandings," 171.

84. Plumwood, "Nature, Self, and Gender," 21.

85. Ibid., 7.

86. Ibid.

87. Slicer, "Your Daughter," 120–21 (emphasis added).

88. Curtin, "Ecological Ethic," 66–68.

89. King, "Caring about Nature," 81, 86.

90. Kheel, "Liberation of Nature," 149.

91. Curtin, "Ecological Ethic," 65.

92. Cheney, "Eco-Feminism," 141.

93. Ibid., 138.

94. Curtin, "Ecological Ethic," 65.

95. Plumwood, "Nature, Self, and Gender," 7.

Chapter 6

1. Samuel Taylor Coleridge, *The Friend,* vol. I, ed. Barbara Rooke (London: Routledge & Kegan Paul, 1969), 514.

2. Michael E. Zimmerman, "Rethinking the Heidegger–Deep Ecology Relationship," *Environmental Ethics* 15 (Fall 1993): 201.

3. Neil Evernden, *The Natural Alien: Humankind and Environment* (Toronto: University of Toronto Press, 1985), 63.

4. George P. Cave, "Animals, Heidegger, and the Right to Life," *Environmental Ethics* 4 (Fall 1982): 251–52.

5. Evernden, *Natural Alien,* 72.

6. Ibid.

7. Ibid., 62, quoting Martin Heidegger, *Being and Time,* trans. John Macquarrie and Edward Robinson (New York: Harper & Row, 1962), 51.

8. John Macquarrie, *Martin Heidegger* (Richmond, VA: John Knox Press, 1968), 11.

9. Evernden, *Natural Alien,* 57.

10. Ibid., 55, 135.

11. Edward C. Relph, "Phenomenology," in *Themes in Geographic Thought,* ed. M. E. Harvey and B. P. Holly (New York: St. Martin's Press, 1981), 112.

12. Ibid., 99, quoting Edmund Husserl, *The Crisis of the European Sciences and Transcendental Phenomenology: An Introduction to Phenomenological Philosophy,* trans. David Carr (Evanston, IL: Northwestern University Press, 1970), 6.

13. Evernden, *Natural Alien,* 58–59.

14. Ibid., 10–11, 22–25, 54, 60, 124.

15. Relph, "Phenomenology," 101.

16. Stephen Fox, *John Muir and His Legacy* (Boston: Little, Brown, 1981), 33, quoting an account published in the *Boston Reporter,* December 21, 1866.

17. Evernden, *Natural Alien,* 9, 19, 25, 33, 50, 55, 70, 77–78, 139, 143.

18. Ibid., 58–59, quoting Erazim Kohak, *Idea and Experience* (Chicago: University of Chicago Press, 1978), 11.

19. Evernden, *Natural Alien,* 80.

20. Relph, "Phenomenology," 102.

21. Evernden, *Natural Alien,* 59.

22. Relph, "Phenomenology," 102, 103, 104.

23. Evernden, *Natural Alien,* 54.

24. Macquarrie, *Martin Heidegger,* 1. Heidegger briefly joined Hitler's National Socialist Party, and although he eventually became disillusioned and withdrew his support, his involvement with the Nazis is a source of much controversy in Heideggerian criticism.

25. George Steiner, *Heidegger* (Hassocks, England: Harvester Press, 1978), 39.

26. Macquarrie, *Martin Heidegger,* 44.

27. Ibid., 39.

28. Evernden, *Natural Alien,* 62, quoting Heidegger, *Being and Time,* 46.

29. Evernden, *Natural Alien,* 60.

30. Macquarrie, *Martin Heidegger,* 45, quoting Heidegger, "What Is Metaphysics," in *Existence and Being,* ed. W. Brock (Chicago: Henry Regnery, 1949).

31. Evernden, *Natural Alien,* 62.

32. Macquarrie, *Martin Heidegger,* 8, quoting Heidegger, *Being and Time,* 236.

33. Macquarrie, *Martin Heidegger,* 10.

34. Evernden, *Natural Alien,* 61.

35. Zimmerman, "Rethinking," 213.

36. Laura Westra, "Let It Be: Heidegger and Future Generations," *Environmental Ethics* 7 (Winter 1985): 345.

37. Martin Heidegger, "The Question Concerning Technology," in *Basic Writings,* ed. David F. Krell (New York: Harper & Row, 1977), 308.

38. Macquarrie, *Martin Heidegger,* 50, quoting Heidegger, *Brief über den Humanismus* (Frankfurt-am-Main: Vittorio Klostermann, 1947), 29.

39. Macquarrie, *Martin Heidegger,* 11.

40. Ibid., 14.

41. Hubert L. Dreyfus, *Being-in-the-World: A Commentary on Heidegger's* Being and Time, *Division 1* (Cambridge: M.I.T. Press, 1991), 149, quoting Heidegger, *Being and Time,* 155.

42. Heidegger, "Question," 313.

43. Hanspeter Padrutt, "Heidegger and Ecology," in *Heidegger and the Earth,* ed. Ladelle McWhorter (Kirksville, MO: Thomas Jefferson University Press, 1992), 11.

44. Dreyfus, *Being-in-the-World,* 149.

45. Cave, "Animals," 251.

46. Dreyfus, *Being-in-the-World,* 61, quoting Heidegger, *Basic Problems of Phenomenology* (Bloomington: Indiana University Press, 1982), 159.

47. Dreyfus, *Being-in-the-World,* 67, quoting Heidegger, *History of the Concept of Time* (Bloomington: Indiana University Press, 1985), 197.

48. Cave, "Animals," 252.

49. Ibid., 253.

50. Dreyfus, *Being-in-the-World,* 131, quoting Heidegger, *Being and Time,* 139.

51. Dreyfus, *Being-in-the-World,* 79, quoting Heidegger, *Being and Time,* 78.

52. Dreyfus, *Being-in-the-World,* 74.

53. Mark Blitz, *Heidegger's* Being and Time *and the Possibility of Political Philosophy* (Ithaca, NY: Cornell University Press, 1981), 168, 67.

54. Relph, "Phenomenology," 109.

55. Dreyfus, *Being-in-the-World,* 90.

56. Dreyfus, *Being-in-the-World,* 165, quoting Heidegger, *Basic Problems of Phenomenology,* 154.

57. Evernden, *Natural Alien,* 63.

58. Dreyfus, *Being-in-the-World,* 133, quoting Heidegger, *Being and Time,* 142.

59. Dreyfus, *Being-in-the-World,* 133.

60. Ibid., 133, quoting Heidegger, *Being and Time,* 142.

61. Evernden, *Natural Alien,* 63.

62. Ibid., 43, 64, quoting William Barrett, *Irrational Man* (Garden City, NY: Anchor, 1962), 217, 218–19.

63. Evernden, *Natural Alien,* 64.

64. Steiner, *Heidegger,* 67.

65. Gail Stenstad, "Singing the Earth," in *Heidegger and the Earth,* ed. Ladelle McWhorter (Kirksville, MO: Thomas Jefferson University Press, 1992), 71.

66. Dreyfus, *Being-in-the-World,* 171, quoting Heidegger, *Being and Time,* 213.

67. Dreyfus, *Being-in-the-World,* 338.

68. Evernden, *Natural Alien,* 128.

69. Heidegger, "Question," 296–98, 309.

70. Macquarrie, *Martin Heidegger,* 46.

71. Bruce Foltz, "On Heidegger and the Interpretation of Environmental Crisis," *Environmental Ethics* 6 (Winter 1984): 328.

72. Eric Lemay and Jennifer A. Pitts, *Heidegger for Beginners* (New York: Writers and Readers, 1994), 71.

73. Steiner, *Heidegger,* 36.

74. Westra, "Let It Be," 348.

75. Foltz, "On Heidegger," 336.

76. Westra, "Let It Be," 348, quoting Heidegger, "Building, Dwelling, Thinking," in *Basic Writings,* ed. David F. Krell (New York: Harper & Row, 1977), 328.

77. Foltz, "On Heidegger," 336.

78. Stenstad, "Singing the Earth," 72.

79. Westra, "Let It Be," 347, quoting Heidegger, "Letter on Humanism," in *Basic Writings,* ed. David F. Krell (New York: Harper & Row, 1977), 196 (emphasis added).

80. Padrutt, "Heidegger and Ecology," 18.

81. Michael E. Zimmerman, "Toward a Heideggerean *Ethos* for Radical Environmentalism," *Environmental Ethics* 5 (Summer 1983): 108.

82. Westra, "Let It Be," 346.

83. Zimmerman, "Toward a Heideggerean *Ethos*," 115, 112.

84. Heidegger, "Question," 310.

85. Zimmerman, "Toward a Heideggerean *Ethos*," 117, quoting Heidegger, "Letter on Humanism," in *Basic Writings*, ed. David F. Krell (New York: Harper & Row, 1977), 210.

86. Foltz, "On Heidegger," 332, quoting Heidegger, *Being and Time*, 96f.

87. Zimmerman, "Toward a Heideggerean *Ethos*," 112.

88. Ibid., 105, quoting Heidegger, *The Question Concerning Technology*, trans. William Lovitt (New York: Harper & Row, 1977), 100.

89. Zimmerman, "Toward a Heideggerean *Ethos*," 102.

90. Westra, "Let It Be," 347.

91. Foltz, "On Heidegger," 331, quoting Heidegger, "Memorial Address," in *Discourse on Thinking*, trans. John M. Anderson and E. Hans Freund (New York: Harper & Row, 1966), 50.

92. Heidegger, "Question," 296–97, 299, 302.

93. Ibid., 309.

94. Dreyfus, *Being-in-the-World*, 231–32, quoting Heidegger, *History of the Concept of Time*, 276.

95. Evernden, *Natural Alien*, 111.

96. Zimmerman, "Toward a Heideggerean *Ethos*," 105.

97. Dreyfus, *Being-in-the-World*, 161.

98. Evernden, *Natural Alien*, 69.

99. Peter Reed, "Man Apart: An Alternative to the Self-Realization Approach," *Environmental Ethics* 11 (Spring 1989): 54.

100. Heidegger, "Question," 310, quoting Friedrich Hölderlin, "Patoms," in *Friedrich Hölderlin Poems and Fragments*, trans. Michael Hamburger (Ann Arbor: University of Michigan Press, 1966), 462–63.

101. Heidegger, "Question," 315.

102. Dreyfus, *Being-in-the-World*, 329.

103. Ibid., 331.

104. Heidegger, "Question," 316.

105. Lemay and Pitts, *Heidegger for Beginners*, 84–85.

106. "Family Room," *Guideposts*, July 1994: 44.

107. Heidegger, "Question," 314, 315, 317 (emphasis added).

108. Dreyfus, *Being-in-the-World*, 339.

109. Ibid.

110. Heidegger, "Question," 317.

111. Stenstad, "Singing the Earth," 74.

112. Evernden, *Natural Alien*, 98.

113. Zimmerman, "Toward a Heideggerean *Ethos*," 119.

114. Ibid., 118, 107–8.

115. Foltz, "On Heidegger," 336–37.

116. Dreyfus, *Being-in-the-World,* 67, quoting Aron Gurwitsch, *Human Encounters in the Social World* (Pittsburgh, PA: Duquesne University Press, 1979), 67.

117. Evernden, *Natural Alien,* 67.

118. Ibid., 48, quoting Marjorie Hope Nicolson, *Mountain Gloom and Mountain Glory: The Development of the Aesthestics of the Infinite* (Ithaca, NY: Cornell University Press, 1959), 1.

119. Evernden, *Natural Alien,* 137.

120. Ibid., 142.

121. Zimmerman, "Toward a Heideggerean *Ethos,*" 123.

Chapter 8

1. The universality of ethical judgment implies that all agents, as such, are understood to be subject to the judgment, not that all agents will as a matter of fact reach the same conclusion; although there is room for moral disagreement in this tradition, there is also recognition that such disagreement frequently results from errors in deliberation among one or all of the parties, especially insofar as the MPV is explicitly designed to yield one correct answer to each dilemma.

Chapter 9

1. Karen Warren, "Feminism and Ecology: Making Connections," *Environmental Ethics* 9 (Spring 1987): 18.

2. Stephen R. Sterling, "Towards an Ecological World View," in *Ethics of Environment and Development: Global Challenge, International Response,* ed. J. Ronald Engel and Joan Gibb Engel (Tucson: University of Arizona Press, 1990), 77–83.

3. Warren, "Feminism and Ecology," 19.

4. Neil Evernden, *The Natural Alien: Humankind and Environment* (Toronto: University of Toronto Press, 1985), 13.

5. Anthony Weston, *Back to Earth: Tomorrow's Environmentalism* (Philadelphia: Temple University Press, 1994), 13, 149–50, 153.

6. Christopher Stone, "Should Trees Have Standing?—Toward Legal Rights for Natural Objects," in *People, Penguins, and Plastic Trees: Basic Issues in Environmental Ethics,* ed. Donald VanDeVeer and Christine Pierce (Belmont CA: Wadsworth Publishing Co., 1986), 83–85, 88.

7. Tom Regan, *The Case for Animal Rights* (Berkeley: University of California Press, 1983), xiii.

8. Paul W. Taylor, *Respect for Nature: A Theory of Environmental Ethics* (Princeton, NJ: Princeton University Press, 1986), 226, 245–46.

9. J. Baird Callicott, *In Defense of the Land Ethic: Essays in Environmental Philosophy* (Albany: State University of New York Press, 1989), 28.

10. Ibid., 3.

11. Laura Westra, "Let It Be: Heidegger and Future Generations," *Environmental Ethics* 7 (Winter 1985): 347–48.

12. Anthony Weston, "Beyond Intrinsic Value: Pragmatism in Environmental Ethics," *Environmental Ethics* 7 (Winter 1985): 328–33.

13. Holmes Rolston, *Philosophy Gone Wild* (Buffalo, NY: Prometheus Books, 1989), 120–21.

14. Aldo Leopold, *A Sand County Almanac, with Essays on Conservation from Round River* (1949; 1953; New York: Ballantine Books, 1966), 42 (emphasis added).

15. Marti Kheel, "The Liberation of Nature: A Circular Affair," *Environmental Ethics* 7 (Summer 1985): 140–41.

16. Eric Katz, "Organism, Community, and the Substitution Problem," in *The Environmental Ethics and Policy Book: Philosophy, Ecology, Economics,* ed. Donald VanDeVeer and Christine Pierce (Belmont, CA: Wadsworth Publishing Co., 1994), 163, 165, 169 (emphasis added).

17. Christopher Stone, "Moral Pluralism and the Course of Environmental Ethics," *Environmental Ethics* 10 (Summer 1988): 142–44, 146, 149, 152.

18. Diana Meyers, "Moral Reflection: Beyond Impartial Reason," *Hypatia* 8.3 (Summer 1993): 29.

19. Kheel, "Liberation of Nature," 138–39.

20. Callicott, *In Defense of the Land,* 46–47, 50–52, 55–59. Callicott draws heavily in this discussion on Mary Midgley's "mixed community" concept as set forth in *Animals and Why They Matter* (Athens: University of Georgia Press, 1983), 112–24.

21. Carol Gilligan, "Moral Orientation and Moral Development," in *Women and Moral Theory,* ed. Eva Feder Kittay and Diana T. Meyers (Totowa, NJ: Rowman & Littlefield, 1987), 20.

22. Lawrence A. Blum, "Gilligan and Kohlberg: Implications for Moral Theory," in *An Ethic of Care: Feminist and Interdisciplinary Perspectives,* ed. Mary Jeanne Larrabee (New York: Routledge, 1993), 50.

23. Gilligan, "Moral Orientation," 20, 25.

24. Meyers, "Moral Reflection," 41.

25. Rita Manning, "Just Caring," in *Explorations in Feminist Ethics,* ed. Eve Browning Cole and Susan Coultrap-McQuin (Bloomington: Indiana University Press, 1992), 50–51, 53.

26. Jim Cheney, "Eco-Feminism and Deep Ecology," *Environmental Ethics* 9 (Summer 1987): 134–45.

27. Graham T. Allison, *Essence of Decision: Explaining the Cuban Missile Crisis* (New York: HarperCollins Publishers, 1971), 258–59.

28. Leopold, *Sand County Almanac,* 3–5, 6–7, 73–74, 116–19, 137–39, 170–71, 212, 251, 263, 265. Students in my Spring 1996 seminar, "Environmental Ethics and Values" (Ecology Curriculum, UNC–Chapel Hill), confirmed my expectation that the analysis of Leopold's writings in this light would prove fruitful and contributed significantly to my thinking along these lines.

Index